Measuring Irrigation Water Flow Rates

Blaine Hanson
Irrigation and Drainage Specialist
Department of Land, Air and Water Resources
University of California, Davis

Larry Schwankl
Irrigation Specialist, UC Kearney Agricultural Center
Department of Land, Air and Water Resources
University of California, Davis

University of California
Agriculture and Natural Resources
Publication 21644

Funded by the Joseph G. Prosser Trust
Administered by the University of California
Water Resources Center

To order or obtain ANR publications and other products, visit the ANR Communication Services online catalog at http://anrcatalog.ucdavis.edu or phone 1-800-994-8849. You can also place orders by mail or FAX, or request a printed catalog of our products from

University of California
Agriculture and Natural Resources
Communication Services
6701 San Pablo Avenue, 2nd Floor
Oakland, California 94608-1239

Telephone 1-800-994-8849
510-642-2431
FAX 510-643-5470
E-mail: danrcs@ucdavis.edu

Publication 21644
ISBN-13: 978-1-60107-642-7

Credits are given in the captions. Cover photo: Lower left, Blaine Hanson; upper left and right, Larry Schwankl. Design by Robin Walton.

This publication has been anonymously peer reviewed for technical accuracy by University of California scientists and other qualified professionals. This review process was managed by the ANR Associate Editor for Land, Air, and Water Sciences.

Printed in USA on recycled paper.

750-pr-11/09-SB/RW

Contents

Preface

This publication is one of a series of water management manuals prepared by the University of California to help California growers address practical irrigation matters. These manuals are designed to provide in-depth information in a semitechnical format. Information on ordering these manuals can be found on page ii.

We thank the Joseph G. Prosser Trust, administered by the University of California Water Resources Center, for the funding to publish this manual. The endowment, established in 1989 by Joseph G. Prosser and his son, Thomas W. Prosser, supports research on the efficient use of irrigation water by researchers in the University of California. In 1951, Thomas Prosser founded Irrometer, Inc. (Riverside, California), an internationally known manufacturer of soil moisture sensors.

Tables

Figures

Why Measure Flow Rates?

Good irrigation water management involves applying the right amount of water at the right time to satisfy crop water use (evapotranspiration). This requires knowing when to irrigate, how much water to apply, and how much water was actually applied. Knowing when to irrigate can be determined from soil moisture measurements or plant-based monitoring; knowing how much water to apply can be determined by keeping track of crop water use through information from the California Irrigation Management Information System (CIMIS).

Knowing how much water was actually applied requires measuring the the flow rate using a flow meter and using the following equation:

$$D = (Q \times T) \div (449 \times A)$$

where
D is the depth of water applied in inches
A is the area irrigated in acres
T is the irrigation time in hours
449 is a conversion factor
Q is the flow rate in gallons per minute (gpm)
T is the time needed to irrigate the entire field if A is the field area; if A is the area per set, T is the time of each irrigation set.

Example

How much water was applied to a 160-acre field using a flow rate of 1,200 gallons per minute and a field irrigation time of 240 hours?

$$D = (1{,}200 \text{ gpm} \times 240 \text{ hours}) \div (449 \times 160 \text{ ac}) = 4.0 \text{ inches}$$

Flow rate measurements can also be used to determine the irrigation time needed to apply a desired amount of water at a given flow rate using the following equation:

$$T = (449 \times D \times A) \div Q$$

Example

What is the irrigation set time needed to apply 3 inches of water over 100 acres at a flow rate of 800 gpm?

$$T = (449 \times 100 \text{ ac} \times 3 \text{ in}) \div 800 \text{ gpm} = 168 \text{ hours}$$

These equations work best for fields irrigated with sprinkle, or microirrigation systems because little surface runoff occurs under these conditions. Thus, the amount applied (D) is the amount that infiltrates the soil. The infiltrated amount can be compared to the crop water use between irrigations to determine whether the irrigation was adequate.

It can be difficult to estimate the amount of water infiltrating the soil under furrow or flood irrigation because *D* includes the surface runoff, and considerable surface runoff can occur. For these irrigation methods, soil moisture monitoring can help determine whether the irrigation was adequate.

This publication provides information for growers and irrigators about devices used to measure flow rates that are particularly appropriate for use on farms. The information includes descriptions of the various flow meters, their installation and operation, and the calculations for determining flow rates and amounts of applied water.

Units of Measurement and Conversion Factors

Volume of Water

The units commonly used for volume measurements in the English system are gallons (gal or g) and cubic feet (ft^3). Units commonly used in the SI (International System of Units), or metric, system are liters (L) and cubic meters (m^3). Because of the large volumes of water used in agriculture, water volumes are normally expressed in acre-feet (ac-ft) or acre-inches (ac-in) in the English system and hectare-meters (ha-m) or hectare-millimeters (ha-mm) in the SI system. To convert these units:

1 cubic foot = 7.48 gallons
1 gallon = 3.785 liters
1 cubic meter = 264 gallons
1 acre-foot = 325,851 gallons = 43,560 cubic feet
1 acre-inch = 27,158 gallons = 3,360 cubic feet
1 acre = 0.40 hectare
1 inch = 2.54 centimeters
1 foot = 0.305 meters

Crop water use, or evapotranspiration, is frequently described as a depth of water that is the volume of water (V) divided by the area irrigated (A), or $V \div A$. The definition of 1 inch of water is 1 acre-inch per acre, or the volume of water ponded 1 inch deep over 1 acre.

Flow Rate

Flow rate is the volume of water per unit of time. English units commonly used for flow rate measurements are gallons per minute (gpm) and cubic feet per second (cfs). In the SI system, the units are liters per second (L/s), liters per minute (L/m), and cubic meters per minute or hour (m^3/s or m^3/h). To convert these units:

1 cubic foot per second = 449 gallons per minute
1 gallon per minute = 3.785 liters per minute
450 gallons per minute = 1 acre-inch per hour

In some areas, the miner's inch is sometimes used for flow rate measurements. A miner's inch is the quantity of water that flows through a 1-inch-square opening under a prescribed head, or level of water above the opening. The head of water can vary from 4 to 6 inches depending on the locality; as a result, the flow rate of a miner's inch, expressed in gallons per minute or cubic feet per second, varies by location. The number of miner's inches is equal to the area of the opening in square inches.

Confusion can occur unless the conditions of the miner's inch are clearly identified. A miner's inch is about 11.25 gallons per minute (with a head of 6 inches) in northern California, Nevada, Arizona, Oregon, and Montana, while it is 9.0 gallons per minute (with a head of 4 inches) in southern California, Idaho, Kansas, Nebraska, South Dakota, New Mexico, Washington, and North Dakota. In general, avoid using the miner's inch as a measurement of flow rate; use meters that measure the rate in gallons per minute, cubic feet per second, or metric units instead.

Flow Rate Measurement Considerations

Many types of flow rate measurement devices are available on the market. However, only a few are particularly suited and commonly used for irrigation systems. The devices used in irrigation systems must be easily installed, easily maintained, and easy to use. The best type of measurement device to use depends on the type of water distribution in the irrigation system. Furrow and flood (border) irrigation systems that use ditches for water distribution need flumes and weirs that can measure open channel flow. Irrigation systems that use pipelines (gated pipe, portable aluminum pipe, or permanent pipelines) need flow meters that can measure pipeline flow.

Open Channel Flow

Open channel flow occurs when the water surface is open to the atmosphere, as occurs in ditches and canals. Atmospheric pressure exists at the water surface. Water flow is caused by elevation differences along the channel, the wetted cross-sectional area, and the shape, length, and roughness of the channel. Usually, elevation differences are due to the slope of the field in the direction of flow. Flumes and weirs are commonly used to measure flow in irrigation ditches, although current velocity meters and the float method can also be used. These latter two, however, require measurements of the cross-sectional area of the ditch.

Pipe Flow

Water flow is caused by pressure differences along the pipeline, which in turn are caused by elevation differences and pressure losses Friction losses are related to the pipe diameter, pipeline length, pipe material, flow rate, and pipe condition.

Many types of flow rate meters can be used in pipelines. They include propeller meters, paddlewheel meters, orifice plates, acoustic flow meters, pitot tube flow meters, vortex flow meters, and electromagnetic flow meters. However, the most commonly used meter for irrigation systems is the propeller meter. Measuring the flow rate of pumping plants frequently involves using the Hall tube or Collins pitot meter.

Flow meters for measuring pipe flow can be classified as velocity averaging meters or point velocity meters. Velocity averaging meters (such as propeller meters) measure the flow rate over a relatively large part of the pipe's cross-sectional area; as a result, the measured flow rate reflects an average velocity across the cross-section. Point velocity meters (such as paddlewheel meters) measure the water velocity at a given point in the pipe's cross-sectional area. Velocity averaging meters are less sensitive to velocity variations across the pipe diameter, while point velocity meters are quite sensitive to this variability.

Water velocity varies across the cross-section of the pipeline. Under good conditions where turbulence in the water is minimal, the water velocity is relatively uniform across the pipe diameter except near the pipe wall, where smaller velocities occur due to friction between the flowing water and the pipe wall (fig. 1).

The water velocity across the pipe diameter can be strongly affected by upstream conditions that cause excessive turbulence in the water. An elbow upstream of the flow meter can cause higher velocities on one side of the pipe (fig. 2). A disc-type check valve can also cause water velocity to vary across the pipe (fig. 3). A partially closed butterfly valve can cause jetting around the edges of the valves, with very high water velocities near the pipe wall and an eddy in the center of the pipe in which water flows back up the pipe, indicated by the negative water velocities in figure 4.

A common recommendation for pipelines is to install flow meters at least 8 to 10 pipeline diameters downstream from a fixture that could cause excessive turbulence in the water. However, research (Hanson and Schwankl 1998b) has shown that velocity averaging flow meters are much less

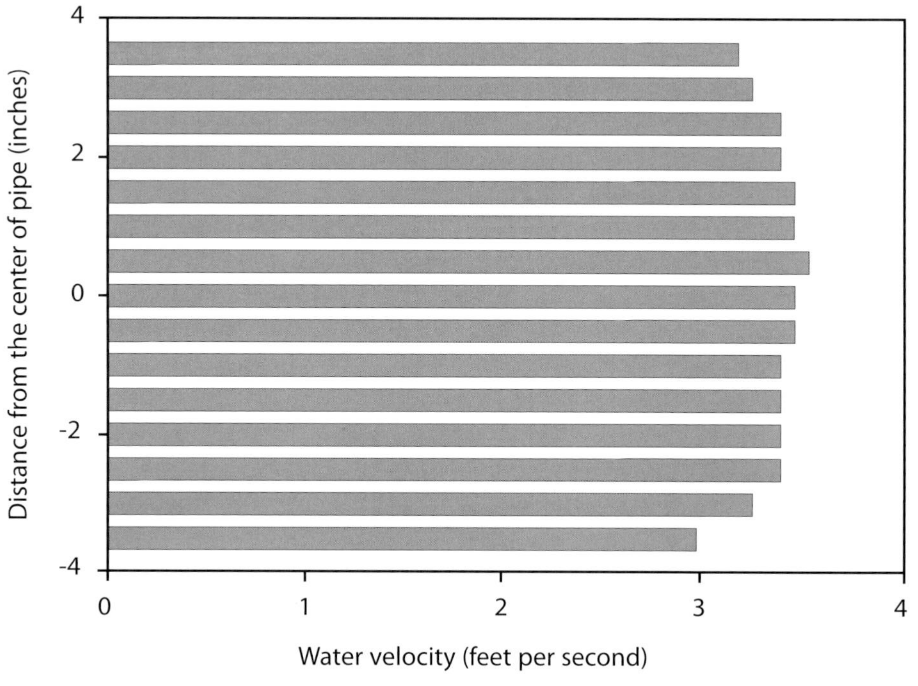

Figure 1. Water velocity distribution under good flow conditions. Pipe diameter is 8 inches.

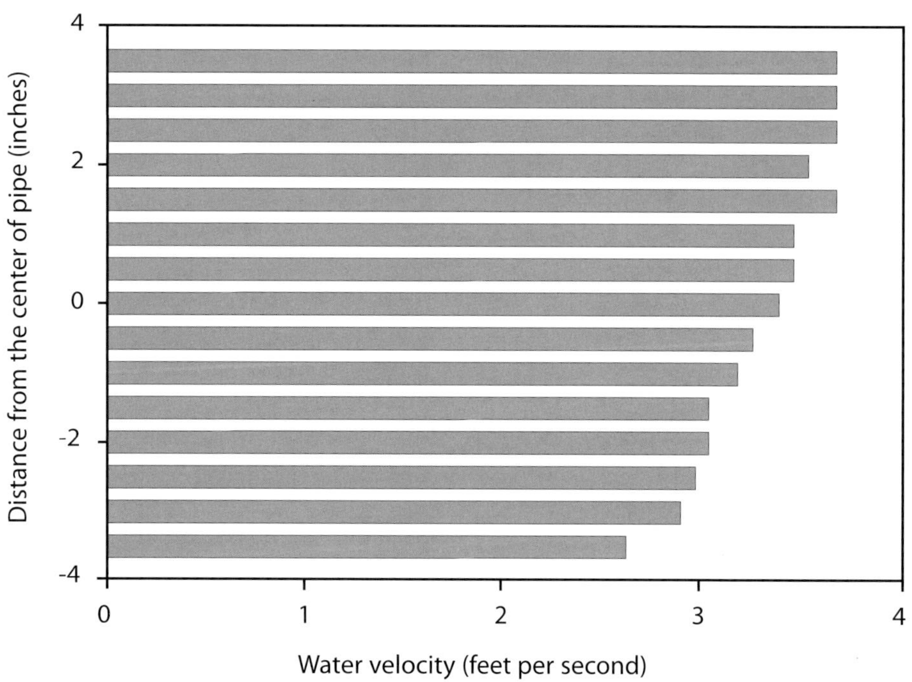

Figure 2. Water velocity distribution immediately downstream from a 90° elbow. Note that the water velocity on one side of the pipeline is about 1.4 times higher than on the other side. Pipe diameter is 8 inches.

affected by turbulence than are point velocity meters, even at 2 pipe diameters downstream. On the other hand, a partially closed butterfly valve can cause large errors in flow rate measurements even at downstream distances exceeding 20 pipe diameters for both types of meters.

Large errors can occur using point velocity flow meters because water velocity can vary across the pipe diameter. A point velocity flow meter installed on the side of the pipe with higher water velocities (see fig. 2) gives a very different flow rate than one installed on the side of the pipe with lower water velocities. Similar behavior could occur because of a check valve (see fig. 3). This behavior can occur at downstream distances exceeding 10 pipe diameters. Use caution when taking point velocity measurements under conditions of excessive turbulence.

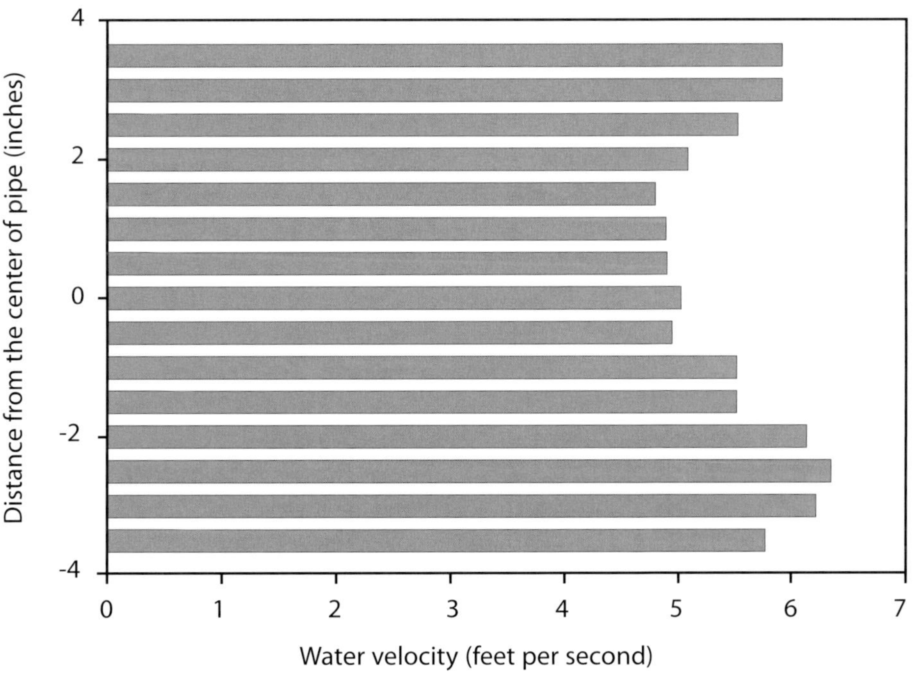

Figure 3. Water velocity distribution immediately downstream from a check valve. Pipe diameter is 8 inches.

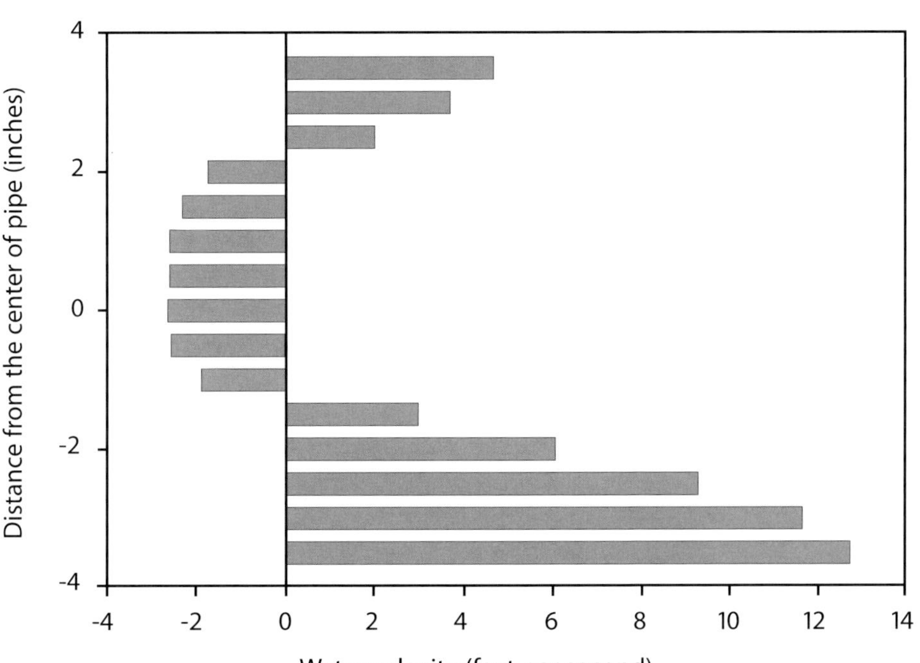

Figure 4. Water velocity distribution immediately downstream of a partially closed butterfly valve. An eddy exists in the center of the pipeline, causing water to flow back up the pipeline (indicated by the negative water velocities). Pipe diameter is 8 inches.

Propeller Flow Meters

Propeller flow meters measure flow rates in fully flowing pipelines, such as those in pressurized irrigation systems. Propeller flow meters respond to an average water velocity across the pipe cross-sectional area because of the diameter of the propeller (and thus are velocity averaging meters). These meters should be used in water that is relatively free of debris.

Description

Propeller flow meters consist of a propeller inserted into a pipeline (fig. 5); a flow rate indicator attached to the meter gives the flow rate, total volume of flow, or both. Most flow rate indicators report in gallons per minute or cubic feet per second, while total flow indicators ("totalizers") report in gallons, acre-feet, or cubic feet. Some indicators report in metric units. Indicators can be dial-type or digital.

Propeller flow meters can be installed by inserting them into a short section of pipe, which is then either coupled (fig. 6A); bolted into the pipeline (fig. 6B); clamped, strapped, or welded onto the pipeline as a saddle-type meter (fig. 6C); or inserted into the pipe discharge (open discharge configuration) (fig 6D).

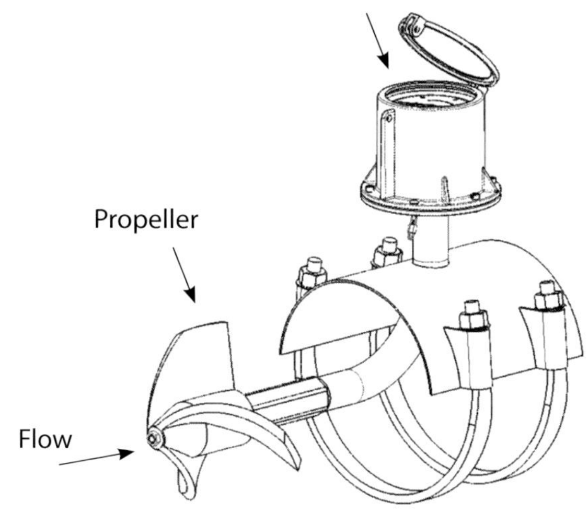

Figure 5. Propeller flow meter. *Source*: Courtesy of McCrometer, Inc., Hemet, CA.

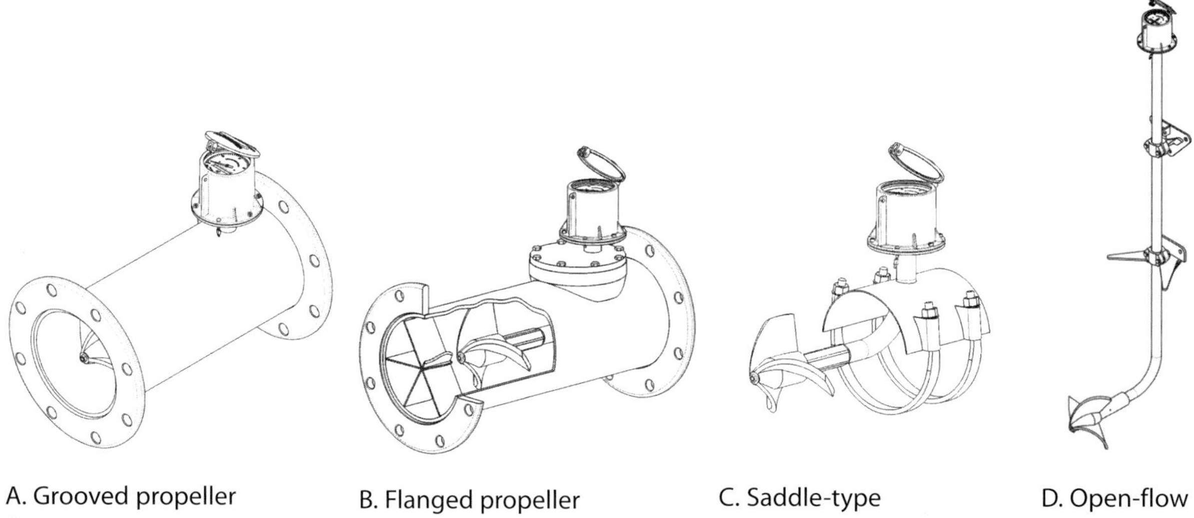

Figure 6. Installation configurations of propeller flow meters. *Source*: Courtesy of McCrometer, Inc., Hemet, CA.

Propeller meters can also be installed as inline meters in a short section of portable pipe. Couplings are welded onto each end of the flow meter to connect it to the portable pipe. One flow meter of a given size can be used for several sizes of portable pipe if a sufficiently long and straight section of pipe of the same diameter as the flow meter section is installed immediately upstream from the flow meter. Hook and latch or ring lock fittings should be used to couple the flow meter section to the pipeline.

Installation

Propeller meters must be matched to the correct pipe size, since the coupling or gear mechanism connecting the propeller to the indicator is based on the pipe diameter and the desired flow rate. Table 1 gives maximum and minimum flow rates for various pipe diameters.

Propeller flow meters operate properly and give accurate readings only if the pipe is flowing full. In pressurized irrigation systems, the flow will usually be full at the pump discharge, but in pumps with an open discharge, as into an irrigation ditch, pipe flow may not be full at the discharge. Care should be used when installing a propeller meter in gated pipe since the pipe may not flow full. The problem can be remedied by creating a slight rise in the discharge pipe; by installing a "gooseneck" at the pipe (fig. 7); by installing an elbow (discharge end pointing upward at the pipe discharge); or by installing a valve downstream from the meter. Propeller flow meters meters can be installed in the horizontal, vertical, or inclined position as long as the pipeline is flowing full.

Table 1. Maximum and minimum flow rates of propeller flow meters for various pipe diameters

	Nominal pipe size (inches)					
	4	**6**	**8**	**10**	**12**	**14**
maximum flow rate (gpm)	600	1,200	1,500	1,800	2,500	3,000
minimum flow rate (gpm)	50	90	100	125	150	250

Source: Courtesy of McCrometer Flowmeters.

The diameter of the propeller may affect the performance of the meter. A propeller whose diameter is large may be more accurate than a propeller with a small diameter because the larger propeller better averages the velocity distribution across the pipeline cross-section.

Propeller flow meters should be installed at a location of minimal water turbulence. The standard recommendation is to install a straight section of pipe, 8 to 10 pipe diameters long, immediately upstream of a propeller flow meter and a straight

Figure 7. Gooseneck added to a pipe to ensure that the pipeline flows full. *Photo*: L. Schwankl.

section about 3 to 5 pipe diameters long immediately downstream of the meter. Research has shown, however, that because of the velocity averaging characteristics of propeller meters, a smaller straight section may be used upstream under some conditions: the error caused by a flap-type check valve 2 pipe diameters upstream from the flow meter was similar that of the control (10 pipe diameters). Flow meter errors were similar to the control error for an elbow installed at 2 to 15 pipe diameters upstream of the meter. However, a partially closed butterfly valve causes significant error in the meter's flow rate measurement, even at 15 pipe diameters upstream.

A relatively stable propeller meter reading means that turbulence is minimal; wide variations in readings signals excessive turbulence. A wildly erratic reading may indicate that air or gas is present in the water.

Straightening Vanes

Straightening vanes, small straight pieces of steel plate attached to the pipe wall, can reduce turbulence in a pipeline. The effectiveness of vanes depends largely on the vane configuration. UC research (Hanson and Schwankl 1998b) showed that one vane attached to the top of the pipe either 2 or 5 pipe diameters downstream from a partially closed butterfly valve was ineffective in reducing the effect of turbulence caused by the valve. However, 6 vanes positioned radially around the pipe circumference at a distance of 2 or 5 pipe diameters downstream of the butterfly valve greatly reduced errors in the flow meter reading compared with a pipe without vanes.

Measuring Flow Rates

Under ideal conditions, a propeller flow meter operated within its recommended range can be accurate to within 2 percent, but the accuracy decreases if the flow rate is too low. Figure 8 shows that accuracy remains within 2 percent as long as the flow rate exceeds about 200 gpm for a 10-inch meter, assuming that no excessive turbulence exists immediately upstream from the meter.

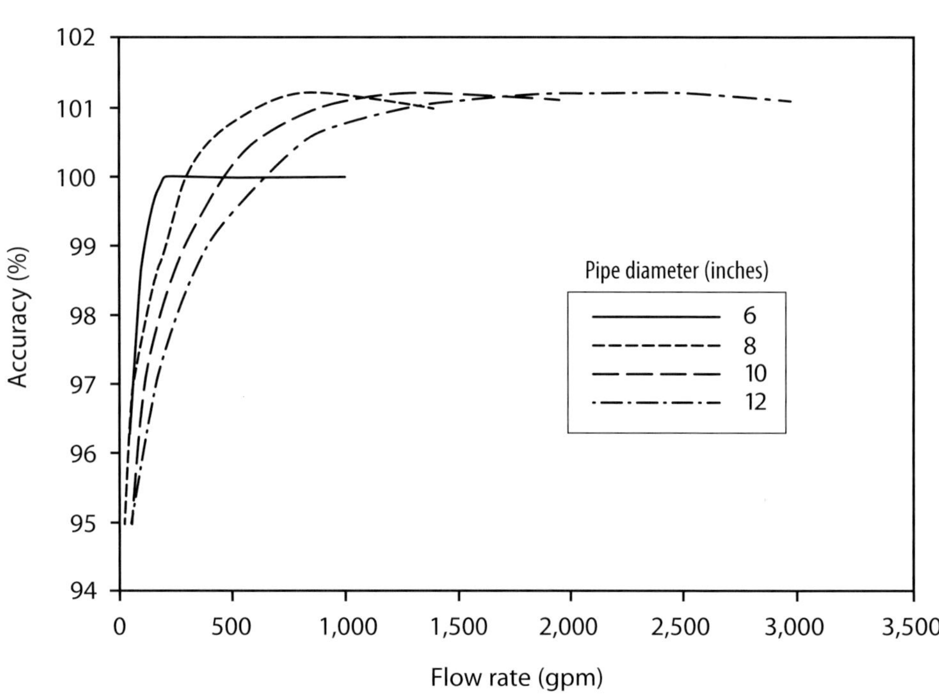

Figure 8. Accuracy of a propeller flow meter for selected pipe diameters and flow rates.

Inserting a propeller into the water flow can cause friction, resulting in pressure, or head, loss across the meter. The amount of pressure loss depends on the velocity or flow rate and the pipe diameter. The higher the flow rate, the more pressure loss because of the flow meter, but the larger the pipe diameter, the smaller the pressure loss. As figure 9 illustrates, the pressure loss caused by propeller meters in a pipeline is generally small. With a 10-inch flow meter, the pressure loss is less than 0.1 psi for flow rates less than 2,000 gpm.

Propeller meters should be calibrated periodically. One large California irrigation district that uses many propeller meters calibrates and maintains them on a 4-year cycle. Wear on the bearings and moving parts can cause deviation from the original calibration. Frequent calibration may be needed for conditions where the irrigation water contains suspended sediment.

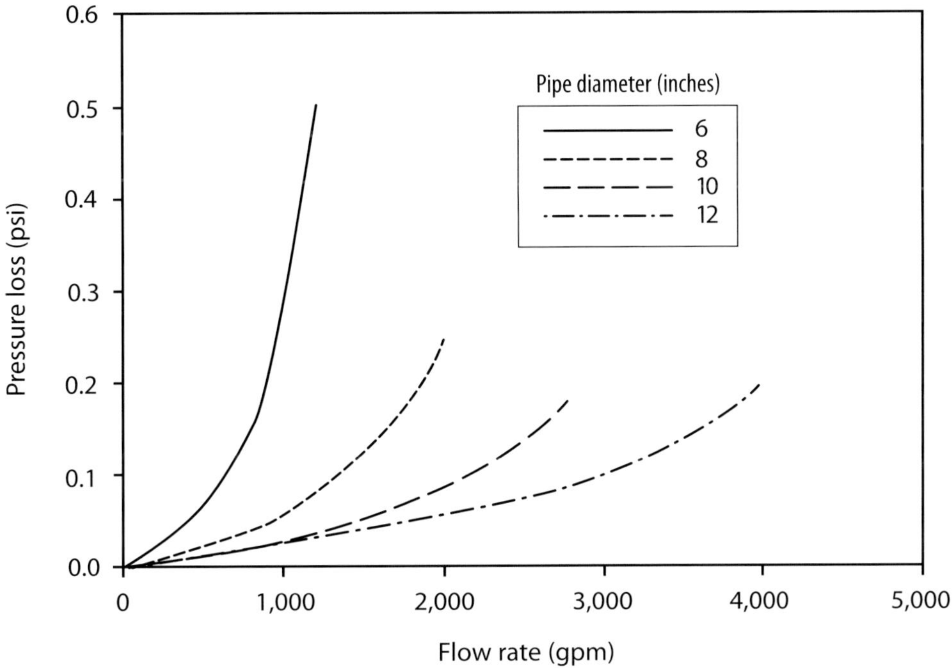

Figure 9. Pressure loss across a propeller flow meter for selected pipe diameters and flow rates.

Considerations

- Flow meters are are best operated near the middle of their design discharge range.
- Avoid spiral flow conditions.
- Avoid installing a meter immediately downstream of a valve that might be partially closed during an irrigation. The jetting action around the valve can cause large errors in the meter's measurement.
- Avoid installing meters in water with weeds and moss, which can foul the propeller and in some cases can stop its rotation (fig. 10).
- Avoid taking measurements when the flow rate is near the minimum recommended flow rate.
- Insert a radial configuration of straightening vanes in the pipeline immediately upstream of the flow meter to remedy conditions of excessive turbulence.
- A centrifugal sand separator or a series of elbows in the pipeline may cause a swirl or rotation to develop in the flowing water. The swirl may still be present even at 100 pipe diameters downstream. The remedy is to place straightening vanes in the pipeline just upstream of the flow meter. A 6-vane straightener with vanes located radially around the pipe diameter is recommended.

Figure 10. Propeller flow meter entangled with weeds and other water contaminants. *Photo:* L. Schwankl.

Electromagnetic Flow Meters

Description

Flow measurement using electromagnetic flow meters, often referred to as "magmeters," is based on Faraday's Law, which states that a conducting material (water in this case) flowing through an electrical field produces a voltage. The magmeter produces an electrical field, measures the induced voltage, and converts it to a flow rate (fig. 11).

Magmeters are very accurate (usually within 1 or 2 percent), place no obstructions in the pipeline, have minimal head loss, and have no moving parts to wear out. While excellent for measuring flow rates in clean water, magmeters have special application in measuring flow rates in water containing debris (e.g., manure waters, canal water with organic matter in it, etc.). Some magmeter manufacturers recommend using less straight pipe upstream and downstream of the meter than that recommended for other pipeline flow meters such as propeller meters. This is because the magmeter's accuracy is less affected by flow turbulence.

Two types of magmeters are available. Tube magmeters (fig. 12) are permanently mounted in the pipeline, most often using flanges. These devices are velocity averaging flow meters. Although most models are heavy, magmeters for pipeline up to 8 or 10 inches can be installed by two people. If installed in an elevated pipeline, especially in PVC pipe, they may need additional support.

Two disadvantages of tube magmeters are that most models require an electrical power supply and that they are more expensive than other meters such as propeller meters. Magmeters are available that can use a range of AC and DC voltages, including battery power. Most tube magmeters are substantially more expensive than propeller meters, with the cost increasing with the size.

Induced voltage = B x D x V
B = Flux density (magnetic strength)
D = Diameter of conductor
V = Fluid mean velocity

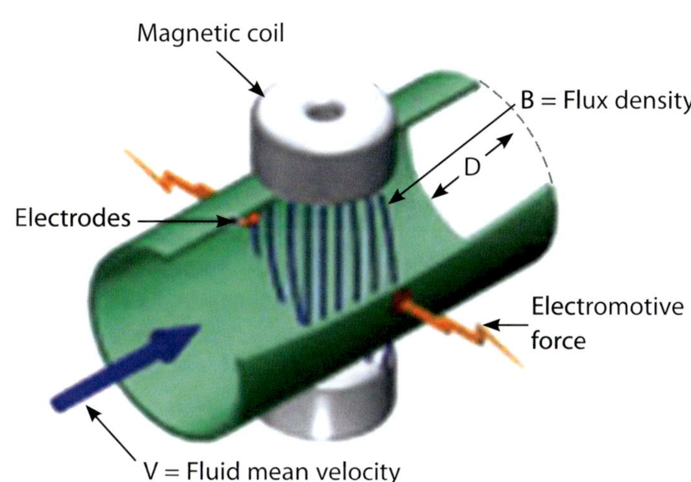

Figure 11. Electromagnetic flow meter. *Source:* Courtesy of McCrometer, Inc., Hemet, CA.

A second type of electromagnetic flow meter is the insertion magmeter, which is inserted through the pipe wall into the flowing water (fig. 13). One type of insertion meter has a single sensor inserted through a threaded tap on the pipe wall and located just on the inside of the pipe wall. While protruding slightly into the pipe, the sensor's streamlined shape is not susceptible to collecting waterborne debris. The same sensor is used no matter what the pipe size, a significant cost savings on larger pipe over the tube magmeters. Another advantage of the insertion magmeter is that it can be moved from one location to another, as long as there is a fitting in the pipeline through which the insertion magmeter can be installed. An insertion magmeter cannot be moved quickly, so its portability is limited.

A disadvantage of insertion magmeters is that they measure the flow velocity only at a single location in the pipeline (point velocity meters). This makes the insertion magmeter more sensitive than a tube magmeter to measurement errors caused by nonuniform flow conditions.

Figure 12. A tube electromagnetic flow meter installed at a dairy to measure manure water flows. *Photo:* L. Schwankl.

Figure 13. Insertion magmeter installed in a PVC pipeline. *Photo:* L. Schwankl.

Pitot Tube Flow Meters

Pitot tube flow meters are velocity averaging meters that measure flow in pressurized pipelines. The pipeline must be flowing full during measurement.

Description

Pitot tube meters consist of a metal tube inserted into the pipeline. The tube has strategically located ports by which the difference in the static and dynamic pressure heads in the water can be measured. (Note that the pressure head in feet equals the pressure in psi divided by 2.31.) The static pressure is simply the pressure in the pipeline, generally measured with a pressure gauge. The dynamic pressure is the pressure of the flowing water due to the water's velocity. The pitot tube is connected to an inverted U-tube water manometer that measures the two pressure heads. One side of the manometer measures the static pressure head, while the other side measures the dynamic pressure head based on the higher water level in that side. This difference between the two pressures is the velocity pressure head. A calibrated sliding scale on the manometer is used along with the pipe diameter to calculate the flow rate.

Pitot tube flow meters are not permanently installed during the irrigation season; they are removed after making a flow rate measurement. They are commonly used to measure the flow rate during a pump test (fig. 14). They are generally not used by irrigators because of the time and expertise required to install and use them and because of the meters' fragility.

Figure 14. Hall pitot tube flow meter installed for pump plant test measurements. *Photo:* B. Hanson.

Types of Pitot Tube Flow Meters

The most common pitot tube flow meters are the Hall (fig. 15) and the Collins (fig. 16), both manufactured by the C. W. Cox Company (http://www.cwcox.com/). The Hall flow meter measures an average water velocity across the pipe diameter through a series of holes on the upstream side of the tube. Tube extensions are provided to adjust the length of the flow meter to fit various pipe diameters.

The Collins pitot tube flow meter is a velocity averaging meter that measures the water velocity at a single point across the pipe diameter. A series of measurements are made at two, six, or ten points across the pipe diameter. The average velocity is calculated from these measurements. A table is provided by the manufacturer to assist setting the flow meter for a desired number of measurements so that each setting represents an equal amount of cross-sectional area.

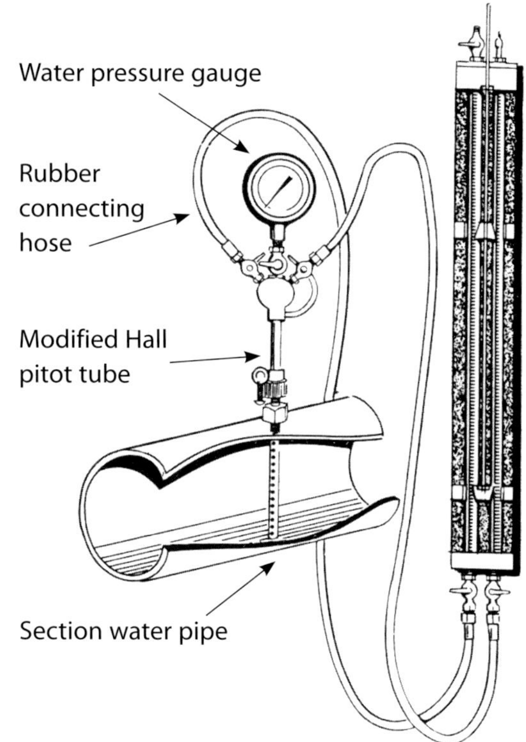

Figure 15. Hall pitot tube flow meter and water manometer. *Source:* Scott and Houston 1977.

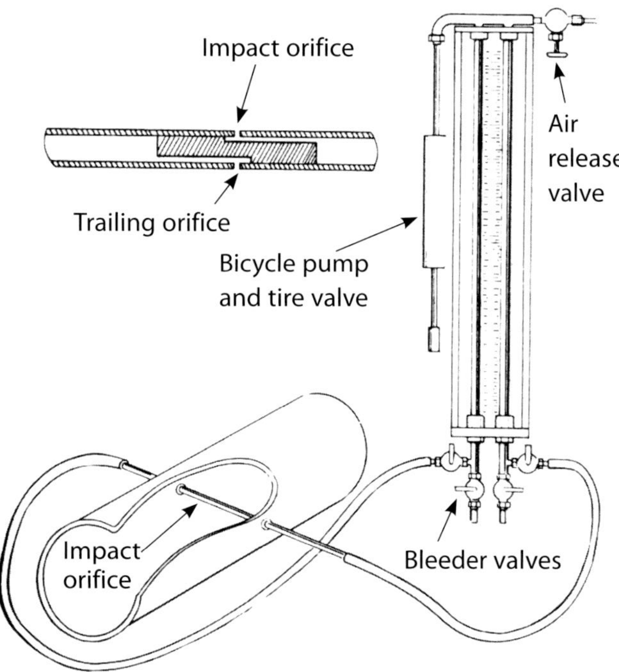

Figure 16. Collins pitot tube flow meter and water manometer. *Source:* Scott and Houston 1977.

These flow meters can be quite accurate: UC research found their accuracy to range from about 1 to 5 percent. A disk-type check valve and a 90° elbow had little effect on these meters at 5 to 15 pipe diameters downstream due to the meters' velocity averaging characteristics.

Considerations

Under high-pressure conditions, the water level on the dynamic pressure head side of a manometer may rise to the top of the manometer and cause water to flow into the static pressure head side, preventing measurement; a bicycle pump is used to pump air into the top of the manometer to push the water levels back down. Also, air or gas in the water can collect in the tubing, affecting manometer readings and preventing accurate measurement. Measurements in an open discharge pipeline, such as when discharging water into a ditch, may not be possible because water may not flow into the manometer due to insufficient pressure (near the open discharge, the water pressure will be close to the atmospheric pressure). In some cases, the manometer could be lowered below the pipe elevation to induce flow into the manometer.

Doppler Flow Meters

Doppler flow meters are velocity averaging meters that are generally used to measure flow rates in pipelines, but they can also be used to measure flow in open channels such as ditches. They have a useful place in the suite of flow meters that should be considered for measuring agricultural pipeline flow rates. Their advantages include true portability, easy use on pipes of various diameters, and operation without obstructing the flow path (they cause no head loss and are not affected by debris or other contaminants in water).

Description

Doppler meters consist of a single (or double) transmitter-receiver unit that is temporarily attached to the outside of a pipeline, along with a display/logger unit (fig. 17). This makes the Doppler meter a truly portable flow meter that can be used on pipes of any diameter.

Measuring Flow Rates

Doppler meters measure the velocity of moving particles in the water by sending out a signal of known frequency and receiving a return signal reflected by the particles. The meter then analyzes any frequency shift in the return signal to determine the particle velocity.

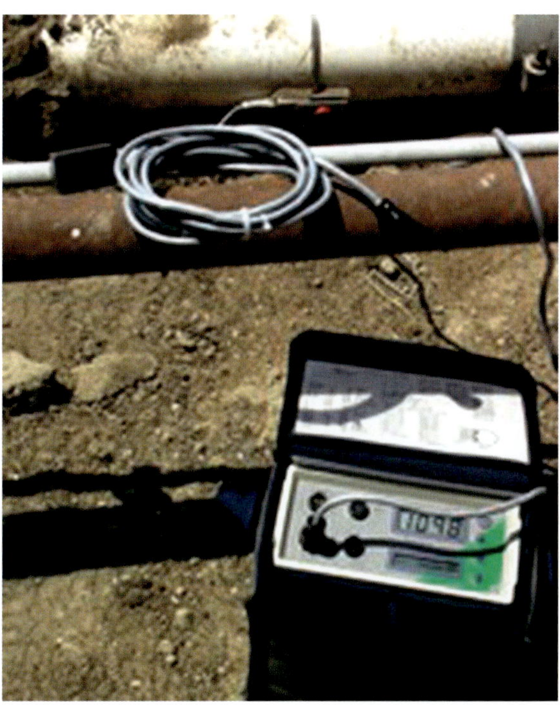

Figure 17. Doppler meter. The transmitter-receiver unit is attached to the pipe and is connected to the battery-powered display/logger unit. *Photo:* L. Schwankl.

The water being measured by a Doppler meter must have impurities in it to allow the meter to work; it will not work in pure or very clean water. Field use has shown that many groundwater sources have enough impurities in them to allow Doppler meters to work, as do the vast majority of surface waters. Waters such as manure water, food processing wastes, and so on also have enough impurities.

Considerations

Doppler meters need an external power source for operation. Both AC and DC units are available, and battery-powered units are available that make the meter more portable and convenient to use.

Doppler meters work well on most metal and plastic pipe but do not work on concrete pipelines. They cost more than propeller meters but less than electromagnetic meters. On large-diameter pipelines, Doppler meters may be very competitively priced.

Accuracy

Doppler meters are not as accurate as electromagnetic or propeller meters. In a test using manure water to compare various magmeters and a calibrated propeller meter with the Doppler meter, the Doppler meter had an error of 5 to 10 percent. This makes it accurate enough for many on-farm applications; when combined with its portability and ease of use, it becomes a useful flow measurement device.

Pipeline flow conditions necessary for accurate Doppler meter flow rate readings are similar to those for propeller meters. There should be 8 to 10 pipe diameters of straight pipe upstream of the Doppler meter and 3 to 5 pipe diameters of straight pipe downstream. Flow conditions downstream of pipeline components (e.g., partially closed valves) that can cause jetting or very nonuniform flow conditions should be avoided.

Trajectory Method

Description

The trajectory method can be used to obtain a rough measurement of the flow rate of water discharged from an open-ended pipe. This method can be used when a rapid estimate of flow rate is needed and a more accurate method is not available. For this method to work, the pipeline must be flowing full.

The trajectory method involves measuring the horizontal distance and the vertical distance of the jet of water discharging from the end of a pipe (fig. 18). The flow rate is obtained by applying these measurements to an appropriate figure or table. The measurements can be made with yardsticks, metersticks, or tape measures.

Measuring Flow Rates

The discharge rate for horizontal flow is calculated using figures 19, 20, and 21 for pipe diameters from 6 to 12 inches. The procedure involves determining the horizontal distance (*X*) from the end of the pipe for a given vertical distance (*Y*) of 6, 12, or 18 inches to the top of the water jet. Accurate measurements can be difficult because very turbulent flow conditions at a distance from the end of the pipe can make it difficult to determine the location of the top of the jet. An error of 10 to 15 percent may exist. Certain conditions are appropriate for this measurement:

- The pipe must be horizontal. For pipe sloping upward, the calculated flow rate will be too high, and for pipes that slope downward, the calculated flow rate will be too low.
- A straight section of pipe at least 6 pipe diameters long should exist upstream from the end of the pipe.
- A constant pipe diameter should exist at least 6 pipe diameters upstream from the end of the pipe.
- The pipe must discharge freely into the air.
- Large fluctuations should not occur in the flow rate, since they may make it difficult to obtain reasonable *X* and *Y* measurements.

For vertical pipe flow, the flow rate is determined by measuring the inside diameter (*D*) of the pipe and the height (*H*) of the jet about the pipe outlet (fig. 22). These measurements are then used with figure 23 to determine the flow rate.

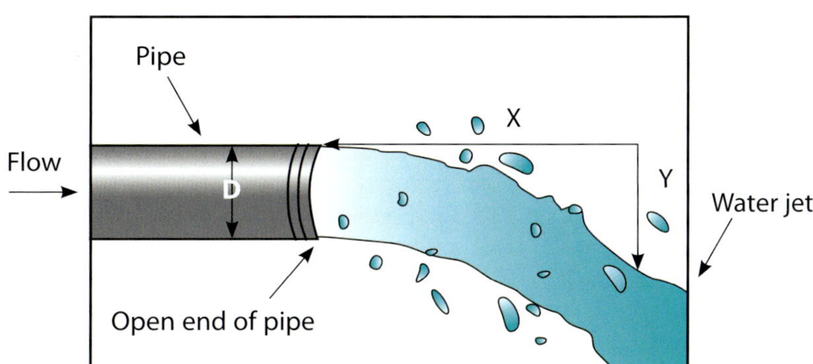

Figure 18. The trajectory method, showing the horizontal distance (X), vertical distance (Y), and the inside diameter of the pipe (D).

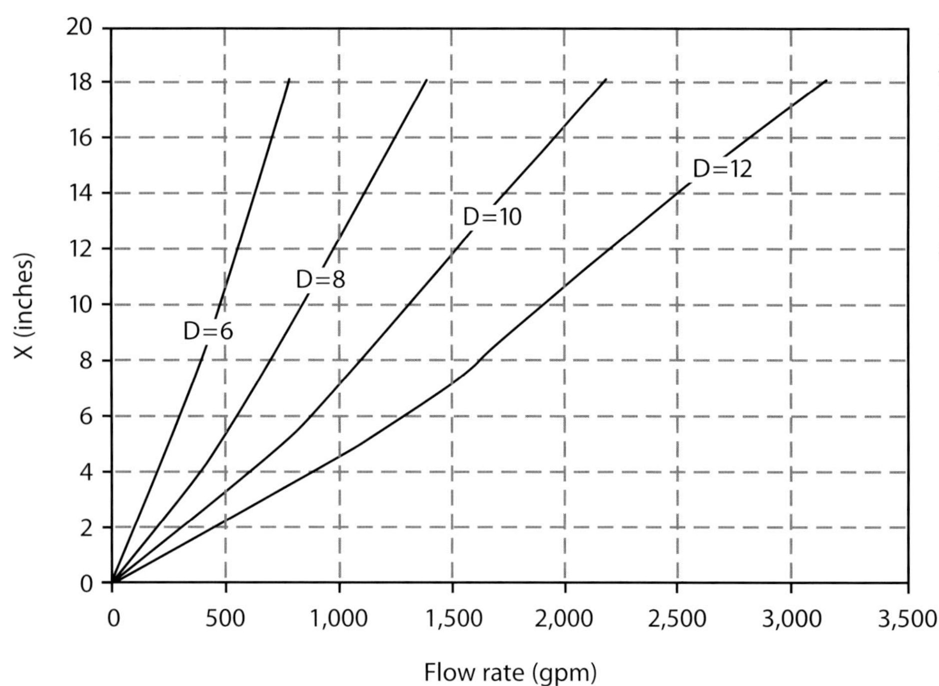

Figure 19. Flow rate chart for vertical distance (Y) = 6 inches and various horizontal distance (X) measurements for pipe diameters (D) of 6, 8, 10, and 12 inches.

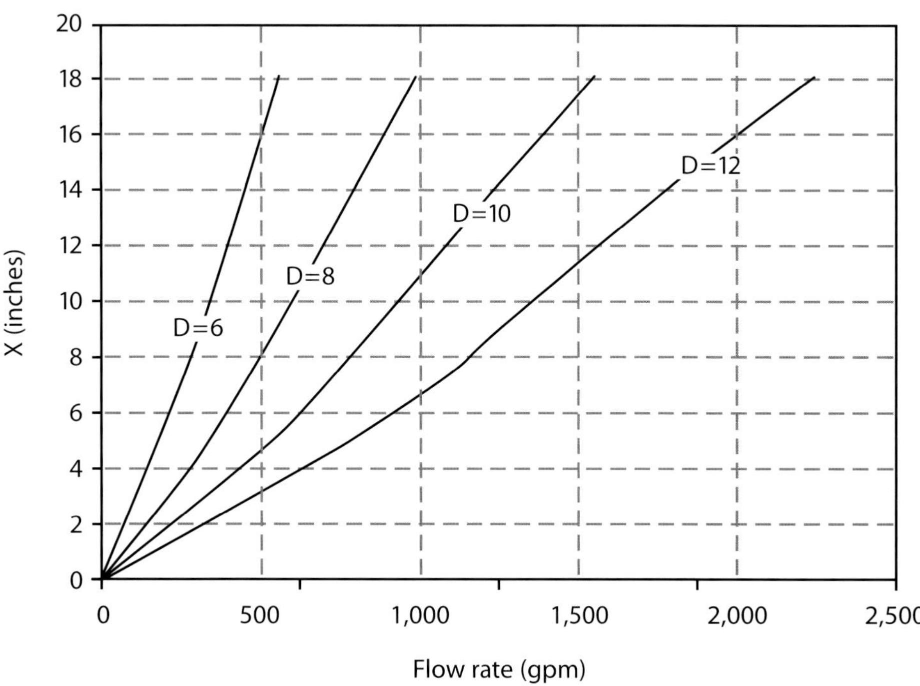

Figure 20. Flow rate chart for vertical distance (Y) = 12 inches and various horizontal distance (X) measurements for pipe diameters (D) of 6, 8, 10, and 12 inches.

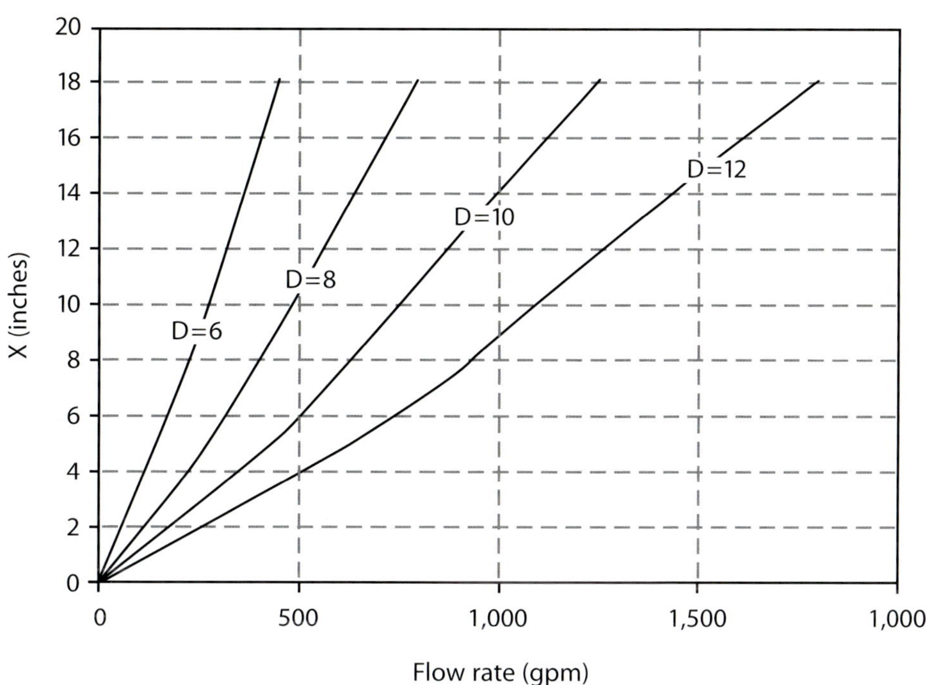

Figure 21. Flow rate chart for a vertical distance (Y) = 18 inches and various horizontal distance (X) measurements for pipe diameters (D) of 6, 8, 10, and 12 inches.

Figure 22. Trajectory method for vertical flow, showing the inside diameter of the pipe (D) and the height of the jet (H).

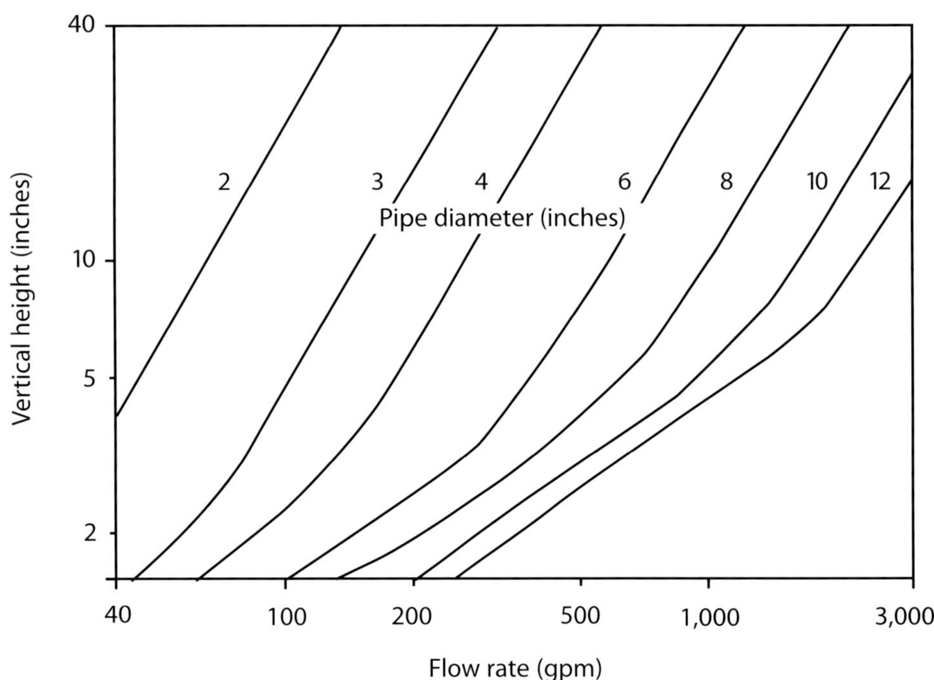

Figure 23. Flow rate chart for vertical flow when using the trajectory method.

California Pipe Method

The California pipe method measures the open-end flow rate of a partially full pipe discharging freely into the air. Measurements required are the inside diameter of the pipe (*D*) and the vertical distance from the top of the pipe to the surface of the flowing water at the end of the pipe (*a*) (fig. 24). These measurements are used with table 2 to calculate the pipe discharge rate. The error in this method may be at least 10 percent. For best results with this method:

- use on pipe diameters of 3 to 10 inches
- the pipe must be partially full
- the ratio of a to *D* must be greater than 0.45
- there must be free discharge at the end of the pipe

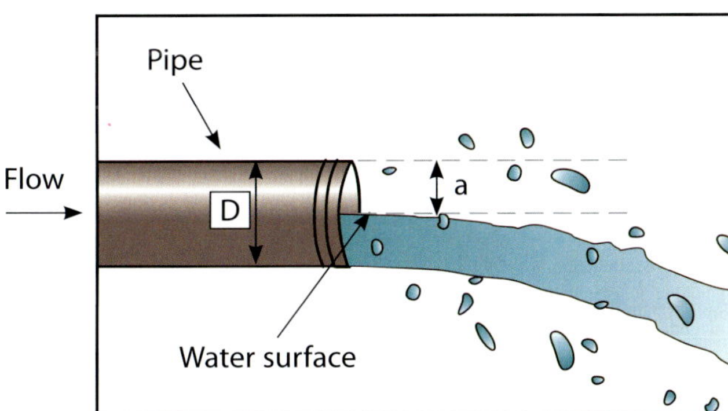

Figure 24. The California pipe method, showing the inside diameter of the pipe (D) and the vertical distance measurement (a).

Table 2. Pipe flow rates using the California pipe method

D (inches)	a/D	Q (cfs)	Q (gpm)	D (inches)	a/D	Q (cfs)	Q (gpm)
4	0.45	0.19	83.2	4	0.75	0.04	18.9
6	0.45	0.51	227.3	6	0.75	0.11	51.6
8	0.45	1.03	463.9	8	0.75	0.23	105.4
10	0.45	1.80	806.8	10	0.75	0.41	183.2
4	0.55	0.13	57.0	4	0.85	0.02	7.2
6	0.55	0.35	155.9	6	0.85	0.04	19.8
8	0.55	0.71	318.1	8	0.85	0.09	40.3
10	0.55	1.23	553.3	10	0.85	0.16	70.1
4	0.65	0.08	35.6	—	—	—	—
6	0.65	0.22	97.2	—	—	—	—
8	0.65	0.44	198.3	—	—	—	—
10	0.65	0.77	344.9	—	—	—	—

Key:
D = inside pipe diameter
a = vertical distance between top of pipe and surface of flowing water
Q = flow rate in cubic feet per second (cfs) or gallons per minute (gpm)

Flumes

Flumes can be used to measure flow rates in ditches. They can be portable for temporary use or permanently installed.

Description

A flume is a specially shaped channel that causes the water flow to accelerate by converging or narrowing the distance between the sidewalls, raising the bottom of the channel, or a combination of both. When the water flows through the constricted section of the flume, a unique relationship exists between flow rate and depth of flow. This flow rate is called critical flow.

There are many types of flumes. This chapter discusses flumes that are particularly suitable for flow measurements on farms. These flumes are commercially available.

Parshall Flume

Parshall flumes (fig. 25) are commonly used to measure flow rate in irrigation ditches, but they are being replaced by flumes with a simpler construction. Parshall flumes are self-cleaning and have a low head loss. Due to their complex construction, they can be difficult to install. They can be permanently installed or portable.

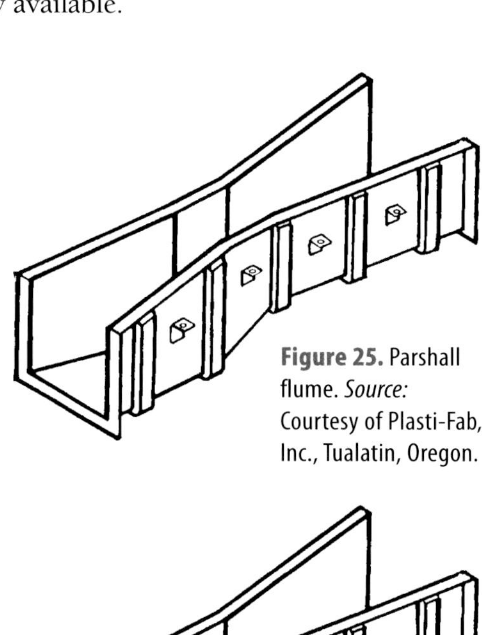

Figure 25. Parshall flume. *Source:* Courtesy of Plasti-Fab, Inc., Tualatin, Oregon.

Cutthroat Flume

The cutthroat flume is a modified Parshall flume with a flat bottom (fig. 26). A disadvantage is that the design may cause variability in calibration.

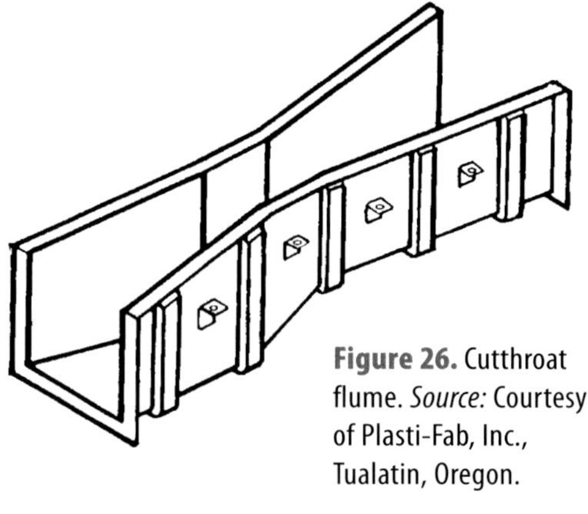

Figure 26. Cutthroat flume. *Source:* Courtesy of Plasti-Fab, Inc., Tualatin, Oregon.

Trapezoidal Flume

Trapezoidal flumes (fig. 27) are trapezoid-shaped with a flat bottom; this configuration more closely conforms to a ditch cross-section than do Parshall or cutthroat flumes. The trapezoidal shape minimizes the length of the upstream transition section of the flow into the flume. It can convey a larger range of flow rates for a given flume and has good accuracy at low flow rates. It can also operate under a higher degree of submergence (80 to 95%) than the Parshall flume without applying corrections for submergence. Portable trapezoidal flumes made of fiberglass are available.

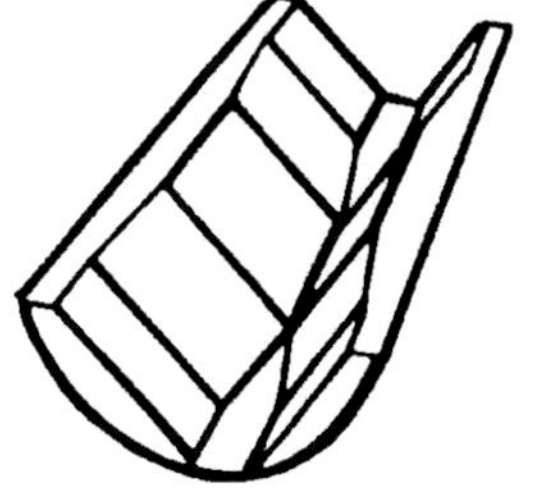

Figure 27. Trapezoidal flume. *Source:* Courtesy of Plasti-Fab, Inc., Tualatin, Oregon.

Broadcrested Weir Flume

The broadcrested weir flume is a trapezoidal flume with a horizontal sill and an upstream ramp leading to the sill (fig. 28). Its trapezoidal shape gives it the same advantages as the trapezoidal flume. It is easily constructed and can be installed in existing concrete-lined ditches. Portable models are available that can be installed in individual furrows. A modified version of this flume is the ramp flume, which has ramps both upstream and downstream of the sill.

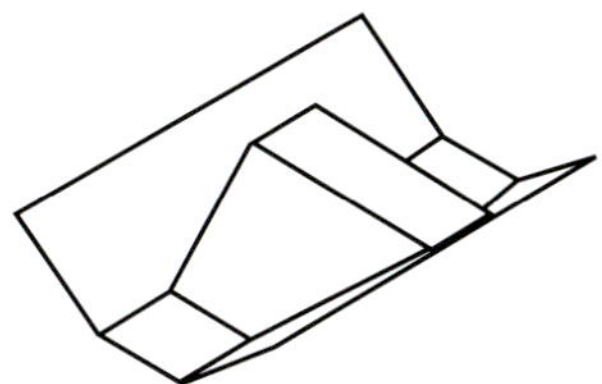

Figure 28. Broadcrested weir flume. *Source:* Courtesy of Plasti-Fab, Inc., Tualatin, Oregon.

Installation

Flumes can be permanently installed in concrete-lined ditches or temporally installed in earth-lined ditches and removed after the irrigation season. Portable flumes made of fiberglass are available from several manufacturers.

The flume must be level along and across the flume. For portable flumes installed in earth-lined ditches, an apron of plastic canvas should be attached to the flume's entrance (fig. 29). The edges of the apron are then buried in the sides and bottom of the ditch. This apron prevents water from flowing along the outside of the flume. Flow around the flume reduces the flow rate in the flume and could eventually wash out the flume. The bottom of the flume should not be below the bottom of the ditch.

Tranquil flow should occur immediately upstream of the flume, defined as the flow that occurs in long, straight channels that have mild slopes and are free of curves, projections, and waves. Conditions of turbulence, surging, unbalanced flow such as would occur around a bend in the ditch, as well as conditions of poorly distributed velocity patterns, should be avoided immediately upstream of the flume to prevent excessive errors. The approach water velocity should exceed 1 foot per second to prevent sediment deposition in the flume.

Figure 29. Trapezoidal flume with an apron attached to the flume inlet. The apron prevents water flowing around the flume. *Photo:* B. Hanson.

The water level downstream of the flume must be low enough to prevent submergence. Submergence occurs when the downstream water level becomes high enough to prevent the acceleration or critical flow of the water. Flow rate readings from a submerged flume will be greatly in error. Submergence can be determined by observing the water flow through the flume. If a frothy wave line exists across the channel below the constricted section, the flume has not reached its limiting submergence. If a wave line does not exist or if the line is at the throat of the flume (within the constricted section), the flume is submerged and the flow rate measurements are invalid. The amount of submergence that can be tolerated by a flume depends to some degree on the type of flume and the downstream conditions. For some flumes, the downstream water level can be measured to adjust the flow rate for submergence. To help reduce submergence problems, the ditch should be free of weeds, grass, trash, and anything else that would impede water flow, and the flume bottom should not be below the bottom of the ditch. A submerged condition can be corrected by raising the flume, which in turn will cause the upstream water level to increase.

Flumes set too low in the ditch may have serious submergence problems. On the other hand, the bottom of the flume should not be too high above the bottom of the ditch. A flume set too high may cause the the water to overflow the ditch bank.

Measuring Flow Rates

Proper operation of a flume causes head loss as the water flows through the flume. Thus, the water level immediately upstream of the flume will be higher than the water level immediately downstream. For a given flow rate, the head loss will depend of the type of flume used. Ditch banks upstream of the flume must be high enough to prevent overtopping, which could be a particular problem for temporarily installed earth-lined ditches commonly used on farms. Generally, the head loss of a flume is about one-quarter of that needed by an equivalent-sized weir for a given flow rate.

The flow rate in a flume is determined by measuring the upstream head at the flume. Calibration equations supplied by the flume manufacturer provide the relationship between head and flow rate.

Measuring the Head

The head of water in the flume is the height of the water level above a reference level. For most flumes, the reference level is the bottom of the flume; for broadcrested weir flumes, the reference level is at the top of the sill. Units of measurement for the head normally are in feet or inches, but they may be in centimeters or meters for SI (International System of Units) units. The following methods can be used to measure the head.

Staff Gauges

Staff gauges installed in the flume's inlet can be read manually (fig. 30). The staff gauge is attached to the side of the flume and must be properly positioned with regard the reference level. The depth marks of the gauge must compensate for the slope of the side of the flume.

Stilling Wells

Stilling wells attached to the flume are frequently used for head measurements. Portable fiberglass flumes can be built with the stilling well attached to the flume as part of the flume (fig. 31). The bottom of the stilling well should be extended by 4 to 6 inches below the bottom of the flume. Before starting the flow, determine the reference level of the flume in the well by filling the well with water until water flows out of the well into the flume; then measure and record the height of the stilling well above the reference level.

Manual measurements of water levels in the stilling well can be made using a tape measure. Measure the height of the top of the stilling well above the bottom of the flume channel or reference level. While the water is flowing, measure the elevation difference between the top of the stilling water and the water level in the well. The depth of the water used for calculating the flow rate is the difference between the height above the reference level and the elevation difference determined during flow measurements. Use the depth of the water in the well with the manufacturer's calibration to calculate the flow rate.

Figure 30. Staff gauge attached to a flume. *Photo:* B. Hanson.

Pressure transducers and data loggers

Continuous measurement of the head is possible by installing pressure transducers in the stilling well and connecting them to data loggers. Several companies make a unit that contains both transducer and data logger. The pressure transducer should compensate for temperature and barometric pressure changes. Manual measurements should be made periodically to check for electronic drift.

The pressure transducer is lowered into the stilling well to a level below the flume channel and remains in place during measurement. The bottom of the stilling well must be below the flume channel. The data logger records the water level above the bottom of the pressure transducer. The initial measurements should be made when there is no flow (fill the stilling well with water until water flows out of the stilling well into the flume) to establish the reference level relative to the pressure transducer depth in the stilling well. The data are downloaded into a computer; the flume head is calculated by subtracting the reference level from the logger readings. The manufacturer's calibration curve is used to determine the flow rates.

Acoustic or laser sensors

Acoustic or laser sensors can be installed above the water level in a stilling well. They measure the elevation difference between the sensor and the water level in the well. The height of the sensor above the flume bottom must be known. The water depth used for flow rate calculations is the difference between the sensor height above the reference level of the flume and the elevation difference between sensor and water level. The calibration of these sensors should be periodically checked. These sensors can be used for continuous measurements of water levels using a data logger.

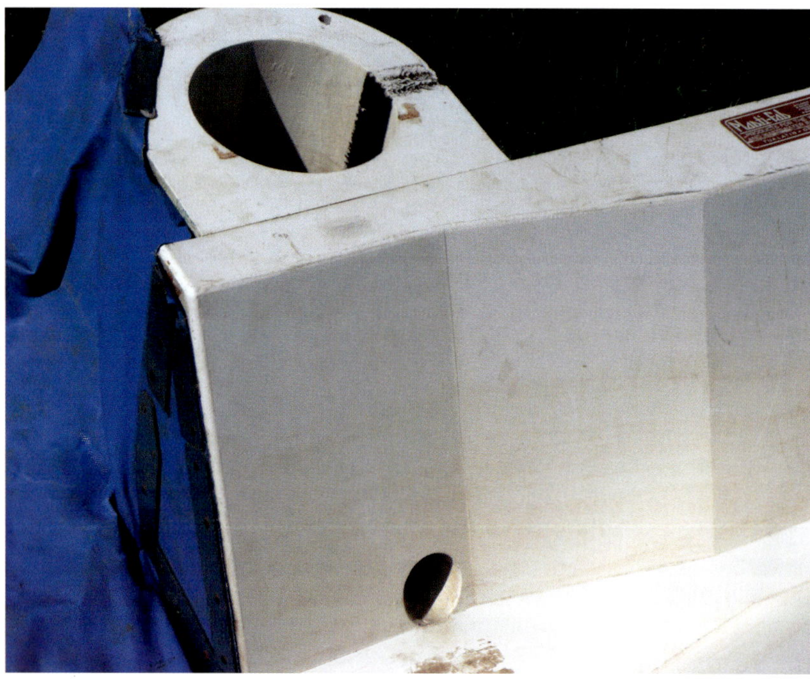

Figure 31. Stilling well attached to the flume. The inlet to the well is located at the bottom of the flume. *Photo:* B. Hanson.

Flume Accuracy

The accuracy of flow measurements from flumes can be affected both by the precision of their construction and by the accuracy with which the head (height of water above a reference point) in the flume is measured. Flumes fabricated in the field are always subject to error in their constructed geometry. Factory-made fiberglass flumes minimize construction error.

Measuring the height of water in a flume is subject to error. The water surface is never completely steady or calm, and even the angle of the observer's eye affects the accuracy.

Follow the installation recommendations closely. Flume accuracy is very sensitive to placement in the channel.

While there may be some error in flume flow measurement, the accuracy is acceptable for most agricultural uses. Of the open-channel flow measurement methods, weirs can be more accurate, but a flume's ability to flush trash and sediment and its small head loss make it a very useful tool for use on the farm.

Flumes can be clogged by weeds, moss, and other floating material in the ditch water, which will affect the accuracy.

Weirs

Description

A weir, a designed notch through which water can flow, can be used to measure flow rate in a ditch or channel (fig. 32). When installed and used properly, they are quite accurate. Weirs are most commonly used in the permanent supply and distribution channels of irrigation districts and suppliers. They are not as commonly used for on-farm water measurements since sediment and trash readily build up behind them and they are difficult to install properly in temporary ditches common on farms.

Weirs are classified according to their notch shape, their configuration relative to the channel, and how water flows across them. The most common weir shapes are triangular, rectangular, and trapezoidal. Weirs are also classified as sharp-crested or broad-crested. Water flowing over a sharp-crested weir does not cling to the weir's downstream face, but creates an air pocket between the water and the weir's downstream face. Broad-crested weirs are less accurate than sharp-crested weirs and are seldom used in irrigation flow measurements.

Water flow through a weir can be contracted or suppressed. Fully contracted flow occurs when the width of the ditch is wider than the width of the notch. Suppressed flow occurs when the sides of the weir coincide with the sides of the ditch. Most weirs used in irrigation flow measurements are contracted. Only fully contracted flow will be discussed here. Detailed information on weirs can be found on the LMNO Engineering, Research, and Software Web site, http:www.lmnoeng.com/Weirs/vweir.htm.

Figure 32. Weirs can be used to measure flow rate. *Photo:* L. Schwankl.

Triangular Weirs

Triangular weirs, often referred to as V-notch weirs, can range in angle from 25º to 100º (fig. 33). The 90º V-notch weir is the most commonly used triangular weir in irrigation. Triangular weirs with a notch angle other than 90º have different calibrations, which can be determined using the equation in the Bureau of Reclamation's *Water Measurement Manual* (2001).

Triangular weirs have a greater range of flow rates and can measure small flow rates with greater accuracy than can other types of weirs because of their triangular shape. However, triangular weirs should not be used for flow rates greater than about 4 cubic feet per minute (1,800 gallons per minute).

Rectangular Weirs

Rectangular weirs (fig. 34) are frequently used due to their ease of construction and relatively good accuracy. They are more suitable for measuring high flow rates than are triangular weirs. Suppressed flow through a rectangular weir occurs when the width of the notch equals the width of channel or ditch.

Trapezoidal Weirs

Trapezoidal weirs are also used for flow measurement. The Cipolletti weir (fig. 35), a trapezoidal weir with notch sides at a horizontal to vertical slope of 1:4, is the most common trapezoidal flume.

Figure 33. Triangular, or V-notch, weir.

Installation

Selecting the Type of Weir to Use

The size and shape of a weir should be based on the channel flow rate and range of flows to be measured. A good estimate of the channel's range of discharge should be made using techniques such as current flow measurement prior to selecting a weir.

Triangular weirs are most often used when it is necessary to measure low flows with high accuracy. Rectangular and trapezoidal weirs have greater capacity but do not measure low flows well.

Figure 34. Rectangular weir.

Figure. 35. Cipolletti wier, a type of trapezoidal weir.

Definitions

The following definitions apply to weirs (fig. 36).

- Crest: The edge or surface that the water passes over.
- Head: The difference between the elevation of the water surface upstream of the weir and the zero reference point.
- Nappe: The sheet of water that passes through the notch and falls over the crest.
- Notch: The shaped opening through which water flows.
- Zero reference point: The lowest elevation of the crest.

Installation Conditions Required for Accurate Measurement

Accurate measurement of weir flow depends on proper installation. The following conditions should be met.

- The weir should be located downstream of a straight channel at least 10 times the length of the weir's crest.
- The weir should be placed at a right angle to the direction of water flow.
- The weir face should be perpendicular; the weir crest must be level.
- The crest and sides of the weir notches should be approximately 1⁄16 inch thick. If the weir plate is thicker than 1⁄16 inch, the downstream edges of the weir crest and sides should be chamfered to an angle of 45° or greater. Knife edges should be avoided, since they can be easily damaged and may be a safety hazard. There should be no nicks in the edges.
- The height of the crest above the channel bottom should be 2 to 3 times the maximum head of the water passing over the weir.
- The distance from the side of the weir to the side of the channel should be at least twice that of the maximum head of water passing over the weir.
- The depth of flow over the weir crest should be at least 2 inches.
- The downstream water level should be at least 2½ inches below the crest; a greater drop is preferred.
- Sediment and debris should not be allowed to build up upstream of the weir. Frequent cleaning may be necessary.
- Conditions of submergence on the downstream side of the weir must be avoided. Submergence occurs when the downstream water level is at or above the elevation of the crest.

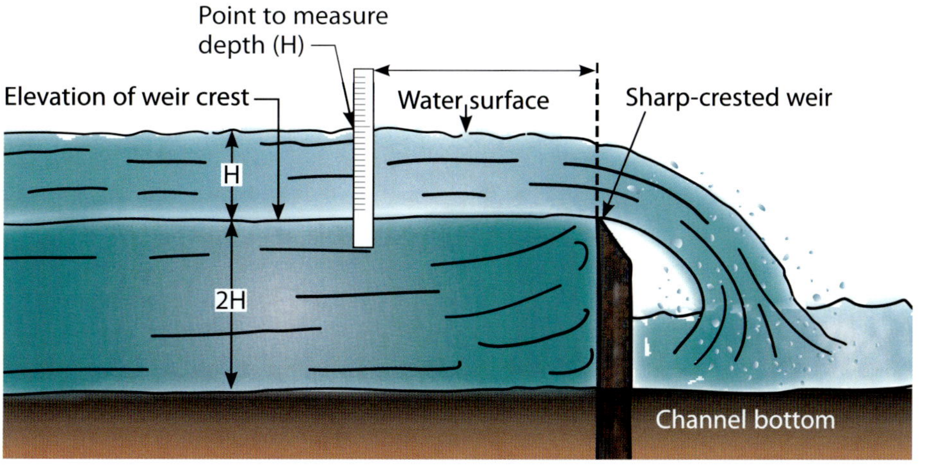

Figure 36. Terms used when measuring flow rate using a weir.

Measuring Flow Rates

Measuring the Flow Depth Over a Weir

As water approaches a weir's crest, there is a "drawdown" as it accelerates to pass through the weir notch (see fig. 36). The water height (head) above the weir crest should be measured upstream where no drawdown occurs. The head of the weir should be measured at an upstream distance equal to 4 to 6 times the maximum head on the crest. The head of the water is the difference between the water surface elevation at this distance and the crest elevation. The crest elevation is the zero reference point. The easiest method of measuring the head is to install a staff gauge upstream from the weir with the zero point at the weir crest elevation. The water depth above the crest can then be easily noted visually. Another option is to permanently set a stake upstream from the weir at the measurement location with its top at the weir crest elevation. The water depth above the end of the stake can then be measured with a tape measure, ruler, or other scale. Note that the head must not be measured at the weir, as this underestimates the flow rate.

Relationships between Flow Rate and Head

The flow rate can be calculated using table 3 for the V-notch weir, table 4 for the rectangular weir, and table 5 for the trapezoidal (Cipolletti) weir. These table values were calculated using the following equations for the respective weir types. Nomenclature for the equations is

Q = flow rate in cubic feet per second (multiply cubic feet per second by 449 to obtain gallons per minute)
H = head of water passing over the weir in feet (see fig. 1)
W = width of crest in feet

V-notch weir (90°)

$$Q = 2.49 \times H^{2.48}$$

Rectangular weir

Contracted weir

$$Q = 3.33 \times H^{1.5} \times (W - 0.2 \times H)$$

Suppressed weir

$$Q = 3.33 \times W \times H^{1.5}$$

Trapezoidal weir (Cipolletti)

$$Q = 3.367 \times W \times H^{1.5}$$

Accuracy

Properly installed weirs are the most accurate open channel flow measurement devices. Measurement error of the flow depth over the weir is a source of inaccuracy, but it can be minimized with care. Weir use is limited for on-farm applications because the weir cannot pass sediment and trash.

Table 3. Relationships between head (H) and flow rate (Q) for a 90° triangular (V-notch) weir

H (feet)	H (inches)	Q (cfs)	Q (gpm)
0.1	1.2	0.008	3.70
0.11	1.32	0.010	4.69
0.12	1.44	0.013	5.82
0.13	1.56	0.016	7.10
0.14	1.68	0.019	8.53
0.15	1.8	0.023	10.12
0.16	1.92	0.026	11.88
0.17	2.04	0.031	13.80
0.18	2.16	0.035	15.90
0.19	2.28	0.041	18.19
0.2	2.4	0.046	20.65
0.21	2.52	0.052	23.31
0.22	2.64	0.058	26.16
0.23	2.76	0.065	29.21
0.24	2.88	0.072	32.46
0.25	3	0.080	35.92
0.26	3.12	0.088	39.59
0.27	3.24	0.097	43.47
0.28	3.36	0.106	47.58
0.29	3.48	0.116	51.90
0.3	3.6	0.126	56.46
0.31	3.72	0.136	61.24
0.32	3.84	0.148	66.25
0.33	3.96	0.159	71.51
0.34	4.08	0.172	77.00
0.35	4.2	0.184	82.74
0.36	4.32	0.198	88.73
0.37	4.44	0.212	94.97
0.38	4.56	0.226	101.46
0.39	4.68	0.241	108.21
0.4	4.8	0.257	115.23
0.41	4.92	0.273	122.50
0.42	5.04	0.290	130.05
0.43	5.16	0.307	137.86
0.44	5.28	0.325	145.95
0.45	5.4	0.344	154.32
0.46	5.52	0.363	162.96
0.47	5.64	0.383	171.89
0.48	5.76	0.403	181.10
0.49	5.88	0.425	190.60
0.5	6	0.446	200.40
0.51	6.12	0.469	210.48
0.52	6.24	0.492	220.87
0.53	6.36	0.516	231.55
0.54	6.48	0.540	242.54
0.55	6.6	0.565	253.83
0.56	6.72	0.591	265.43
0.57	6.84	0.618	277.34
0.58	6.96	0.645	289.57
0.59	7.08	0.673	302.11
0.6	7.2	0.701	314.96
0.61	7.32	0.731	328.14
0.62	7.44	0.761	341.65
0.63	7.56	0.792	355.48
0.64	7.68	0.823	369.63
0.65	7.8	0.856	384.12
0.66	7.92	0.889	398.95
0.67	8.04	0.922	414.11
0.68	8.16	0.957	429.60
0.69	8.28	0.992	445.44
0.7	8.4	1.028	461.62
0.71	8.52	1.065	478.15
0.72	8.64	1.103	495.03
0.73	8.76	1.141	512.26
0.74	8.88	1.180	529.83
0.75	9	1.220	547.77
0.76	9.12	1.261	566.06
0.77	9.24	1.302	584.71
0.78	9.36	1.345	603.73
0.79	9.48	1.388	623.10

H (feet)	H (inches)	Q (cfs)	Q (gpm)
0.8	9.6	1.432	643
0.81	9.72	1.477	663
0.82	9.84	1.522	683
0.83	9.96	1.569	704
0.84	10.08	1.616	726
0.85	10.2	1.664	747
0.86	10.32	1.713	769
0.87	10.44	1.763	792
0.88	10.56	1.813	814
0.89	10.68	1.865	837
0.9	10.8	1.917	861
0.91	10.92	1.971	885
0.92	11.04	2.025	909
0.93	11.16	2.080	934
0.94	11.28	2.136	959
0.95	11.4	2.193	984
0.96	11.52	2.250	1010
0.97	11.64	2.309	1037
0.98	11.76	2.368	1063
0.99	11.88	2.429	1090
1	12	2.490	1118
1.01	12.12	2.552	1146
1.02	12.24	2.615	1174
1.03	12.36	2.679	1203
1.04	12.48	2.744	1232
1.05	12.6	2.810	1262
1.06	12.72	2.877	1292
1.07	12.84	2.945	1322
1.08	12.96	3.014	1353
1.09	13.08	3.083	1384
1.1	13.2	3.154	1416
1.11	13.32	3.226	1448
1.12	13.44	3.298	1481
1.13	13.56	3.372	1514
1.14	13.68	3.446	1547
1.15	13.8	3.522	1581
1.16	13.92	3.598	1615
1.17	14.04	3.675	1650
1.18	14.16	3.754	1685
1.19	14.28	3.833	1721
1.2	14.4	3.914	1757
1.21	14.52	3.995	1794
1.22	14.64	4.077	1831
1.23	14.76	4.161	1868
1.24	14.88	4.245	1906
1.25	15	4.330	1944
1.26	15.12	4.417	1983
1.27	15.24	4.504	2022
1.28	15.36	4.593	2062
1.29	15.48	4.682	2102
1.3	15.6	4.773	2143
1.31	15.72	4.864	2184
1.32	15.84	4.957	2226
1.33	15.96	5.051	2268
1.34	16.08	5.145	2310
1.35	16.2	5.241	2353
1.36	16.32	5.338	2397
1.37	16.44	5.436	2441
1.38	16.56	5.535	2485
1.39	16.68	5.635	2530
1.4	16.8	5.736	2575
1.41	16.92	5.838	2621
1.42	17.04	5.941	2668
1.43	17.16	6.046	2714
1.44	17.28	6.151	2762
1.45	17.4	6.257	2810
1.46	17.52	6.365	2858
1.47	17.64	6.474	2907
1.48	17.76	6.583	2956
1.49	17.88	6.694	3006
1.5	18	6.806	3056

Table 4. Relationships between head (H) and flow rate (Q) for a rectangular weir for various crest widths

H (feet)	H (inches)	Crest width (feet)									
		1		1.5		2		3		4	
		Q (cfs)	Q (gpm)	Q (cfs)	Q (gpm)	Q (cfs)	Q (gpm)	Q (cfs)	Q (gpm)	Q (cfs)	Q (gpm)
0.1	1.2	0.103	46	0.156	70	0.209	94	0.314	141	0.419	188
0.11	1.32	0.119	53	0.180	81	0.240	108	0.362	162	0.483	217
0.12	1.44	0.135	61	0.204	92	0.274	123	0.412	185	0.550	247
0.13	1.56	0.152	68	0.230	103	0.308	138	0.464	208	0.620	279
0.14	1.68	0.170	76	0.257	115	0.344	154	0.518	233	0.693	311
0.15	1.8	0.188	84	0.284	128	0.381	171	0.575	258	0.768	345
0.16	1.92	0.206	93	0.313	140	0.419	188	0.633	284	0.846	380
0.17	2.04	0.225	101	0.342	154	0.459	206	0.692	311	0.926	416
0.18	2.16	0.245	110	0.372	167	0.499	224	0.754	338	1.008	453
0.19	2.28	0.265	119	0.403	181	0.541	243	0.817	367	1.093	491
0.2	2.4	0.286	128	0.435	195	0.584	262	0.882	396	1.179	530
0.21	2.52	0.307	138	0.467	210	0.627	282	0.948	426	1.268	570
0.22	2.64	0.329	147	0.500	225	0.672	302	1.016	456	1.359	610
0.23	2.76	0.350	157	0.534	240	0.718	322	1.085	487	1.452	652
0.24	2.88	0.373	167	0.568	255	0.764	343	1.156	519	1.547	695
0.25	3	0.395	178	0.604	271	0.812	364	1.228	551	1.644	738
0.26	3.12	0.419	188	0.639	287	0.860	386	1.301	584	1.743	783
0.27	3.24	0.442	198	0.676	303	0.909	408	1.376	618	1.844	828
0.28	3.36	0.466	209	0.712	320	0.959	431	1.453	652	1.946	874
0.29	3.48	0.490	220	0.750	337	1.010	453	1.530	687	2.050	920
0.3	3.6	0.514	231	0.788	354	1.062	477	1.609	722	2.156	968
0.31	3.72	0.539	242	0.827	371	1.114	500	1.689	758	2.263	1016
0.32	3.84	0.564	253	0.866	389	1.167	524	1.770	795	2.373	1065
0.33	3.96	0.590	265	0.905	406	1.221	548	1.852	832	2.483	1115
0.34	4.08	0.615	276	0.945	424	1.275	573	1.936	869	2.596	1166
0.35	4.2	0.641	288	0.986	443	1.331	598	2.020	907	2.710	1217
0.36	4.32	0.667	300	1.027	461	1.387	623	2.106	946	2.825	1269
0.37	4.44	0.694	312	1.069	480	1.443	648	2.193	985	2.942	1321
0.38	4.56	0.721	324	1.111	499	1.501	674	2.281	1024	3.061	1374
0.39	4.68	0.748	336	1.153	518	1.559	700	2.370	1064	3.181	1428
0.4	4.8	0.775	348	1.196	537	1.617	726	2.460	1104	3.302	1483
0.41	4.92	0.803	360	1.240	557	1.677	753	2.551	1145	3.425	1538
0.42	5.04	0.830	373	1.283	576	1.737	780	2.643	1187	3.549	1594
0.43	5.16	0.858	385	1.328	596	1.797	807	2.736	1229	3.675	1650
0.44	5.28	0.886	398	1.372	616	1.858	834	2.830	1271	3.802	1707
0.45	5.4	0.915	411	1.417	636	1.920	862	2.925	1313	3.930	1765
0.46	5.52	0.943	424	1.463	657	1.982	890	3.021	1357	4.060	1823
0.47	5.64	0.972	436	1.509	677	2.045	918	3.118	1400	4.191	1882
0.48	5.76	1.001	449	1.555	698	2.108	947	3.216	1444	4.323	1941
0.49	5.88	1.030	463	1.601	719	2.172	975	3.315	1488	4.457	2001
0.5	6	1.060	476	1.648	740	2.237	1004	3.414	1533	4.592	2062
0.51	6.12	1.089	489	1.696	761	2.302	1034	3.515	1578	4.728	2123
0.52	6.24	1.119	502	1.743	783	2.367	1063	3.616	1624	4.865	2184
0.53	6.36	1.149	516	1.791	804	2.434	1093	3.718	1670	5.003	2246
0.54	6.48	1.179	529	1.839	826	2.500	1123	3.821	1716	5.143	2309
0.55	6.6	1.209	543	1.888	848	2.567	1153	3.925	1763	5.284	2372
0.56	6.72	1.239	556	1.937	870	2.635	1183	4.030	1810	5.426	2436
0.57	6.84	1.270	570	1.986	892	2.703	1214	4.136	1857	5.569	2500

Table 4, cont.

H (feet)	H (inches)	Crest width (feet) 1		1.5		2		3		4	
		Q (cfs)	Q (gpm)	Q (cfs)	Q (gpm)	Q (cfs)	Q (gpm)	Q (cfs)	Q (gpm)	Q (cfs)	Q (gpm)
0.58	6.96	1.300	584	2.036	914	2.771	1244	4.242	1905	5.713	2565
0.59	7.08	1.331	598	2.086	936	2.840	1275	4.349	1953	5.858	2630
0.6	7.2	1.362	612	2.136	959	2.910	1306	4.457	2001	6.005	2696
0.61	7.32	1.393	625	2.186	982	2.979	1338	4.566	2050	6.152	2762
0.62	7.44	1.424	639	2.237	1004	3.050	1369	4.675	2099	6.301	2829
0.63	7.56	1.455	653	2.288	1027	3.121	1401	4.786	2149	6.451	2896
0.64	7.68	1.487	668	2.339	1050	3.192	1433	4.897	2199	6.602	2964
0.65	7.8	1.518	682	2.391	1073	3.263	1465	5.008	2249	6.753	3032
0.66	7.92	1.550	696	2.443	1097	3.335	1498	5.121	2299	6.906	3101
0.67	8.04	1.582	710	2.495	1120	3.408	1530	5.234	2350	7.060	3170
0.68	8.16	1.613	724	2.547	1144	3.481	1563	5.348	2401	7.215	3240
0.69	8.28	1.645	739	2.600	1167	3.554	1596	5.462	2453	7.371	3310
0.7	8.4	1.677	753	2.652	1191	3.627	1629	5.578	2504	7.528	3380
0.71	8.52	1.709	767	2.705	1215	3.701	1662	5.694	2556	7.686	3451
0.72	8.64	1.741	782	2.759	1239	3.776	1695	5.810	2609	7.845	3522
0.73	8.76	1.774	796	2.812	1263	3.851	1729	5.928	2662	8.005	3594
0.74	8.88	1.806	811	2.866	1287	3.926	1763	6.046	2714	8.165	3666
0.75	9	1.838	825	2.920	1311	4.001	1797	6.164	2768	8.327	3739
0.76	9.12	1.871	840	2.974	1335	4.077	1831	6.284	2821	8.490	3812
0.77	9.24	1.903	855	3.028	1360	4.153	1865	6.403	2875	8.653	3885
0.78	9.36	1.936	869	3.083	1384	4.230	1899	6.524	2929	8.818	3959
0.79	9.48	1.969	884	3.138	1409	4.307	1934	6.645	2984	8.983	4034
0.8	9.6	2.002	899	3.193	1434	4.384	1969	6.767	3038	9.150	4108
0.81	9.72	2.034	913	3.248	1458	4.462	2003	6.889	3093	9.317	4183
0.82	9.84	2.067	928	3.303	1483	4.540	2038	7.012	3149	9.485	4259
0.83	9.96	2.100	943	3.359	1508	4.618	2074	7.136	3204	9.654	4335
0.84	10.08	2.133	958	3.415	1533	4.697	2109	7.260	3260	9.824	4411
0.85	10.2	2.166	973	3.471	1558	4.776	2144	7.385	3316	9.995	4488
0.86	10.32	2.199	987	3.527	1584	4.855	2180	7.511	3372	10.166	4565
0.87	10.44	2.232	1002	3.583	1609	4.934	2215	7.637	3429	10.339	4642
0.88	10.56	2.265	1017	3.640	1634	5.014	2251	7.763	3486	10.512	4720
0.89	10.68	2.298	1032	3.696	1660	5.094	2287	7.890	3543	10.686	4798
0.9	10.8	2.331	1047	3.753	1685	5.175	2323	8.018	3600	10.861	4877
0.91	10.92	2.365	1062	3.810	1711	5.255	2360	8.146	3658	11.037	4956
0.92	11.04	2.398	1077	3.867	1736	5.336	2396	8.275	3715	11.213	5035
0.93	11.16	2.431	1092	3.924	1762	5.418	2432	8.404	3773	11.391	5114
0.94	11.28	2.464	1106	3.982	1788	5.499	2469	8.534	3832	11.569	5194
0.95	11.4	2.498	1121	4.039	1814	5.581	2506	8.664	3890	11.748	5275
0.96	11.52	2.531	1136	4.097	1840	5.663	2543	8.795	3949	11.927	5355
0.97	11.64	2.564	1151	4.155	1865	5.745	2580	8.927	4008	12.108	5436
0.98	11.76	2.597	1166	4.213	1892	5.828	2617	9.059	4067	12.289	5518
0.99	11.88	2.631	1181	4.271	1918	5.911	2654	9.191	4127	12.471	5600
1	12	2.664	1196	4.329	1944	5.994	2691	9.324	4186	12.654	5682
1.01	12.12	2.697	1211	4.387	1970	6.077	2729	9.457	4246	12.838	5764
1.02	12.24	2.731	1226	4.446	1996	6.161	2766	9.591	4307	13.022	5847
1.03	12.36	2.764	1241	4.504	2022	6.245	2804	9.726	4367	13.207	5930
1.04	12.48	2.797	1256	4.563	2049	6.329	2842	9.861	4427	13.393	6013
1.05	12.6	2.830	1271	4.622	2075	6.413	2880	9.996	4488	13.579	6097
1.06	12.72	2.864	1286	4.681	2102	6.498	2918	10.132	4549	13.766	6181

Table 4, cont.

H (feet)	H (inches)	Crest width (feet) 1 Q (cfs)	1 Q (gpm)	1.5 Q (cfs)	1.5 Q (gpm)	2 Q (cfs)	2 Q (gpm)	3 Q (cfs)	3 Q (gpm)	4 Q (cfs)	4 Q (gpm)
1.07	12.84	2.897	1301	4.740	2128	6.583	2956	10.268	4610	13.954	6265
1.08	12.96	2.930	1316	4.799	2155	6.668	2994	10.405	4672	14.143	6350
1.09	13.08	2.963	1331	4.858	2181	6.753	3032	10.542	4734	14.332	6435
1.1	13.2	2.997	1345	4.917	2208	6.838	3070	10.680	4795	14.522	6520
1.11	13.32	3.030	1360	4.977	2235	6.924	3109	10.818	4857	14.713	6606
1.12	13.44	3.063	1375	5.036	2261	7.010	3147	10.957	4920	14.904	6692
1.13	13.56	3.096	1390	5.096	2288	7.096	3186	11.096	4982	15.096	6778
1.14	13.68	3.129	1405	5.156	2315	7.182	3225	11.236	5045	15.289	6865
1.15	13.8	3.162	1420	5.215	2342	7.269	3264	11.376	5108	15.482	6952
1.16	13.92	3.195	1435	5.275	2369	7.356	3303	11.516	5171	15.676	7039
1.17	14.04	3.228	1449	5.335	2396	7.442	3342	11.657	5234	15.871	7126
1.18	14.16	3.261	1464	5.395	2422	7.529	3381	11.798	5297	16.066	7214
1.19	14.28	3.294	1479	5.455	2449	7.617	3420	11.940	5361	16.262	7302
1.2	14.4	3.327	1494	5.516	2476	7.704	3459	12.082	5425	16.459	7390
1.21	14.52	3.360	1508	5.576	2504	7.792	3499	12.224	5489	16.656	7479
1.22	14.64	3.392	1523	5.636	2531	7.880	3538	12.367	5553	16.854	7568
1.23	14.76	3.425	1538	5.696	2558	7.968	3577	12.510	5617	17.053	7657
1.24	14.88	3.458	1553	5.757	2585	8.056	3617	12.654	5682	17.252	7746
1.25	15	3.490	1567	5.817	2612	8.144	3657	12.798	5746	17.452	7836
1.26	15.12	3.523	1582	5.878	2639	8.233	3696	12.942	5811	17.652	7926
1.27	15.24	3.555	1596	5.938	2666	8.321	3736	13.087	5876	17.853	8016
1.28	15.36	3.588	1611	5.999	2694	8.410	3776	13.233	5941	18.055	8107
1.29	15.48	3.620	1625	6.060	2721	8.499	3816	13.378	6007	18.257	8197
1.3	15.6	3.653	1640	6.120	2748	8.588	3856	13.524	6072	18.460	8289
1.31	15.72	3.685	1654	6.181	2775	8.678	3896	13.671	6138	18.663	8380
1.32	15.84	3.717	1669	6.242	2803	8.767	3936	13.817	6204	18.867	8471
1.33	15.96	3.749	1683	6.303	2830	8.857	3977	13.964	6270	19.072	8563
1.34	16.08	3.781	1698	6.364	2857	8.946	4017	14.112	6336	19.277	8655
1.35	16.2	3.813	1712	6.425	2885	9.036	4057	14.260	6403	19.483	8748
1.36	16.32	3.845	1726	6.486	2912	9.126	4098	14.408	6469	19.689	8840
1.37	16.44	3.877	1741	6.547	2939	9.216	4138	14.556	6536	19.896	8933
1.38	16.56	3.908	1755	6.608	2967	9.307	4179	14.705	6603	20.104	9026
1.39	16.68	3.940	1769	6.669	2994	9.397	4219	14.854	6670	20.312	9120
1.4	16.8	3.972	1783	6.730	3022	9.488	4260	15.004	6737	20.520	9214
1.41	16.92	4.003	1797	6.791	3049	9.578	4301	15.154	6804	20.729	9307
1.42	17.04	4.035	1811	6.852	3076	9.669	4342	15.304	6872	20.939	9402
1.43	17.16	4.066	1826	6.913	3104	9.760	4382	15.455	6939	21.149	9496
1.44	17.28	4.097	1840	6.974	3131	9.851	4423	15.605	7007	21.360	9591
1.45	17.4	4.128	1854	7.035	3159	9.942	4464	15.757	7075	21.571	9685
1.46	17.52	4.159	1867	7.096	3186	10.034	4505	15.908	7143	21.783	9780
1.47	17.64	4.190	1881	7.158	3214	10.125	4546	16.060	7211	21.995	9876
1.48	17.76	4.221	1895	7.219	3241	10.217	4587	16.212	7279	22.208	9971
1.49	17.88	4.252	1909	7.280	3269	10.308	4628	16.365	7348	22.421	10067
1.5	18	4.282	1923	7.341	3296	10.400	4670	16.518	7416	22.635	10163

Table 5. Relationships between head (H) and flow rate (Q) for a trapezoidal (Cipoletti) weir for various crest widths

H (feet)	H (inches)	Crest width (feet)									
		1		1.5		2		3		4	
		Q (cfs)	Q (gpm)	Q (cfs)	Q (gpm)	Q (cfs)	Q (gpm)	Q (cfs)	Q (gpm)	Q (cfs)	Q (gpm)
0.1	1.2	0.106	48	0.160	72	0.213	96	0.319	143	0.426	191
0.11	1.32	0.123	55	0.184	83	0.246	110	0.369	165	0.491	221
0.12	1.44	0.140	63	0.210	94	0.280	126	0.420	189	0.560	251
0.13	1.56	0.158	71	0.237	106	0.316	142	0.473	213	0.631	283
0.14	1.68	0.176	79	0.265	119	0.353	158	0.529	238	0.705	317
0.15	1.8	0.196	88	0.293	132	0.391	176	0.587	263	0.782	351
0.16	1.92	0.215	97	0.323	145	0.431	194	0.646	290	0.862	387
0.17	2.04	0.236	106	0.354	159	0.472	212	0.708	318	0.944	424
0.18	2.16	0.257	115	0.386	173	0.514	231	0.771	346	1.029	462
0.19	2.28	0.279	125	0.418	188	0.558	250	0.837	376	1.115	501
0.2	2.4	0.301	135	0.452	203	0.602	270	0.903	406	1.205	541
0.21	2.52	0.324	145	0.486	218	0.648	291	0.972	436	1.296	582
0.22	2.64	0.347	156	0.521	234	0.695	312	1.042	468	1.390	624
0.23	2.76	0.371	167	0.557	250	0.743	334	1.114	500	1.486	667
0.24	2.88	0.396	178	0.594	267	0.792	355	1.188	533	1.584	711
0.25	3	0.421	189	0.631	283	0.842	378	1.263	567	1.684	756
0.26	3.12	0.446	200	0.670	301	0.893	401	1.339	601	1.786	802
0.27	3.24	0.472	212	0.709	318	0.945	424	1.417	636	1.890	848
0.28	3.36	0.499	224	0.748	336	0.998	448	1.497	672	1.995	896
0.29	3.48	0.526	236	0.789	354	1.052	472	1.577	708	2.103	944
0.3	3.6	0.553	248	0.830	373	1.107	497	1.660	745	2.213	994
0.31	3.72	0.581	261	0.872	391	1.162	522	1.743	783	2.325	1044
0.32	3.84	0.609	274	0.914	410	1.219	547	1.828	821	2.438	1095
0.33	3.96	0.638	287	0.957	430	1.277	573	1.915	860	2.553	1146
0.34	4.08	0.668	300	1.001	450	1.335	599	2.003	899	2.670	1199
0.35	4.2	0.697	313	1.046	470	1.394	626	2.092	939	2.789	1252
0.36	4.32	0.727	327	1.091	490	1.455	653	2.182	980	2.909	1306
0.37	4.44	0.758	340	1.137	510	1.516	680	2.273	1021	3.031	1361
0.38	4.56	0.789	354	1.183	531	1.577	708	2.366	1062	3.155	1417
0.39	4.68	0.820	368	1.230	552	1.640	736	2.460	1105	3.280	1473
0.4	4.8	0.852	382	1.278	574	1.704	765	2.555	1147	3.407	1530
0.41	4.92	0.884	397	1.326	595	1.768	794	2.652	1191	3.536	1588
0.42	5.04	0.916	411	1.375	617	1.833	823	2.749	1234	3.666	1646
0.43	5.16	0.949	426	1.424	639	1.899	853	2.848	1279	3.798	1705
0.44	5.28	0.983	441	1.474	662	1.965	882	2.948	1324	3.931	1765
0.45	5.4	1.016	456	1.525	685	2.033	913	3.049	1369	4.066	1825
0.46	5.52	1.050	472	1.576	707	2.101	943	3.151	1415	4.202	1887
0.47	5.64	1.085	487	1.627	731	2.170	974	3.255	1461	4.340	1948
0.48	5.76	1.120	503	1.680	754	2.239	1005	3.359	1508	4.479	2011
0.49	5.88	1.155	519	1.732	778	2.310	1037	3.465	1556	4.620	2074
0.5	6	1.190	534	1.786	802	2.381	1069	3.571	1603	4.762	2138
0.51	6.12	1.226	551	1.839	826	2.453	1101	3.679	1652	4.905	2202
0.52	6.24	1.263	567	1.894	850	2.525	1134	3.788	1701	5.050	2268
0.53	6.36	1.299	583	1.949	875	2.598	1167	3.897	1750	5.197	2333
0.54	6.48	1.336	600	2.004	900	2.672	1200	4.008	1800	5.344	2400
0.55	6.6	1.373	617	2.060	925	2.747	1233	4.120	1850	5.493	2467
0.56	6.72	1.411	634	2.116	950	2.822	1267	4.233	1901	5.644	2534

Table 5, cont.

H (feet)	H (inches)	Crest width (feet) 1 Q (cfs)	1 Q (gpm)	1.5 Q (cfs)	1.5 Q (gpm)	2 Q (cfs)	2 Q (gpm)	3 Q (cfs)	3 Q (gpm)	4 Q (cfs)	4 Q (gpm)
0.57	6.84	1.449	651	2.173	976	2.898	1301	4.347	1952	5.796	2602
0.58	6.96	1.487	668	2.231	1002	2.975	1336	4.462	2003	5.949	2671
0.59	7.08	1.526	685	2.289	1028	3.052	1370	4.578	2055	6.104	2740
0.6	7.2	1.565	703	2.347	1054	3.130	1405	4.695	2108	6.259	2810
0.61	7.32	1.604	720	2.406	1080	3.208	1441	4.812	2161	6.416	2881
0.62	7.44	1.644	738	2.466	1107	3.287	1476	4.931	2214	6.575	2952
0.63	7.56	1.684	756	2.525	1134	3.367	1512	5.051	2268	6.735	3024
0.64	7.68	1.724	774	2.586	1161	3.448	1548	5.172	2322	6.896	3096
0.65	7.8	1.764	792	2.647	1188	3.529	1584	5.293	2377	7.058	3169
0.66	7.92	1.805	811	2.708	1216	3.611	1621	5.416	2432	7.221	3242
0.67	8.04	1.847	829	2.770	1244	3.693	1658	5.540	2487	7.386	3316
0.68	8.16	1.888	848	2.832	1272	3.776	1695	5.664	2543	7.552	3391
0.69	8.28	1.930	866	2.895	1300	3.860	1733	5.789	2599	7.719	3466
0.7	8.4	1.972	885	2.958	1328	3.944	1771	5.916	2656	7.888	3542
0.71	8.52	2.014	904	3.021	1357	4.029	1809	6.043	2713	8.057	3618
0.72	8.64	2.057	924	3.086	1385	4.114	1847	6.171	2771	8.228	3694
0.73	8.76	2.100	943	3.150	1414	4.200	1886	6.300	2829	8.400	3772
0.74	8.88	2.143	962	3.215	1444	4.287	1925	6.430	2887	8.573	3849
0.75	9	2.187	982	3.280	1473	4.374	1964	6.561	2946	8.748	3928
0.76	9.12	2.231	1002	3.346	1502	4.462	2003	6.692	3005	8.923	4007
0.77	9.24	2.275	1021	3.412	1532	4.550	2043	6.825	3064	9.100	4086
0.78	9.36	2.319	1041	3.479	1562	4.639	2083	6.958	3124	9.278	4166
0.79	9.48	2.364	1062	3.546	1592	4.728	2123	7.093	3185	9.457	4246
0.8	9.6	2.409	1082	3.614	1623	4.818	2163	7.228	3245	9.637	4327
0.81	9.72	2.455	1102	3.682	1653	4.909	2204	7.364	3306	9.818	4408
0.82	9.84	2.500	1123	3.750	1684	5.000	2245	7.500	3368	10.001	4490
0.83	9.96	2.546	1143	3.819	1715	5.092	2286	7.638	3429	10.184	4573
0.84	10.08	2.592	1164	3.888	1746	5.184	2328	7.776	3492	10.369	4656
0.85	10.2	2.639	1185	3.958	1777	5.277	2369	7.916	3554	10.554	4739
0.86	10.32	2.685	1206	4.028	1809	5.371	2411	8.056	3617	10.741	4823
0.87	10.44	2.732	1227	4.098	1840	5.465	2454	8.197	3680	10.929	4907
0.88	10.56	2.780	1248	4.169	1872	5.559	2496	8.339	3744	11.118	4992
0.89	10.68	2.827	1269	4.241	1904	5.654	2539	8.481	3808	11.308	5077
0.9	10.8	2.875	1291	4.312	1936	5.750	2582	8.624	3872	11.499	5163
0.91	10.92	2.923	1312	4.384	1969	5.846	2625	8.769	3937	11.691	5249
0.92	11.04	2.971	1334	4.457	2001	5.942	2668	8.913	4002	11.885	5336
0.93	11.16	3.020	1356	4.530	2034	6.039	2712	9.059	4068	12.079	5423
0.94	11.28	3.069	1378	4.603	2067	6.137	2756	9.206	4133	12.274	5511
0.95	11.4	3.118	1400	4.676	2100	6.235	2800	9.353	4199	12.471	5599
0.96	11.52	3.167	1422	4.751	2133	6.334	2844	9.501	4266	12.668	5688
0.97	11.64	3.217	1444	4.825	2166	6.433	2889	9.650	4333	12.867	5777
0.98	11.76	3.266	1467	4.900	2200	6.533	2933	9.799	4400	13.066	5867
0.99	11.88	3.317	1489	4.975	2234	6.633	2978	9.950	4467	13.266	5957
1	12	3.367	1512	5.051	2268	6.734	3024	10.101	4535	13.468	6047
1.01	12.12	3.418	1535	5.126	2302	6.835	3069	10.253	4604	13.671	6138
1.02	12.24	3.469	1557	5.203	2336	6.937	3115	10.406	4672	13.874	6229
1.03	12.36	3.520	1580	5.279	2370	7.039	3161	10.559	4741	14.079	6321

Table 5, cont.

H (feet)	H (inches)	Crest width (feet) 1 Q (cfs)	1 Q (gpm)	1.5 Q (cfs)	1.5 Q (gpm)	2 Q (cfs)	2 Q (gpm)	3 Q (cfs)	3 Q (gpm)	4 Q (cfs)	4 Q (gpm)
1.04	12.48	3.571	1603	5.357	2405	7.142	3207	10.713	4810	14.284	6414
1.05	12.6	3.623	1627	5.434	2440	7.245	3253	10.868	4880	14.491	6506
1.06	12.72	3.675	1650	5.512	2475	7.349	3300	11.024	4950	14.698	6599
1.07	12.84	3.727	1673	5.590	2510	7.453	3347	11.180	5020	14.907	6693
1.08	12.96	3.779	1697	5.669	2545	7.558	3394	11.337	5090	15.116	6787
1.09	13.08	3.832	1720	5.747	2581	7.663	3441	11.495	5161	15.326	6882
1.1	13.2	3.884	1744	5.827	2616	7.769	3488	11.653	5232	15.538	6977
1.11	13.32	3.938	1768	5.906	2652	7.875	3536	11.813	5304	15.750	7072
1.12	13.44	3.991	1792	5.986	2688	7.982	3584	11.973	5376	15.964	7168
1.13	13.56	4.044	1816	6.067	2724	8.089	3632	12.133	5448	16.178	7264
1.14	13.68	4.098	1840	6.147	2760	8.197	3680	12.295	5520	16.393	7360
1.15	13.8	4.152	1864	6.228	2797	8.305	3729	12.457	5593	16.609	7458
1.16	13.92	4.207	1889	6.310	2833	8.413	3778	12.620	5666	16.826	7555
1.17	14.04	4.261	1913	6.392	2870	8.522	3826	12.783	5740	17.044	7653
1.18	14.16	4.316	1938	6.474	2907	8.632	3876	12.948	5813	17.263	7751
1.19	14.28	4.371	1963	6.556	2944	8.742	3925	13.112	5888	17.483	7850
1.2	14.4	4.426	1987	6.639	2981	8.852	3975	13.278	5962	17.704	7949
1.21	14.52	4.481	2012	6.722	3018	8.963	4024	13.444	6037	17.926	8049
1.22	14.64	4.537	2037	6.806	3056	9.074	4074	13.611	6112	18.149	8149
1.23	14.76	4.593	2062	6.890	3093	9.186	4125	13.779	6187	18.372	8249
1.24	14.88	4.649	2087	6.974	3131	9.298	4175	13.948	6262	18.597	8350
1.25	15	4.706	2113	7.058	3169	9.411	4226	14.117	6338	18.822	8451
1.26	15.12	4.762	2138	7.143	3207	9.524	4276	14.286	6415	19.048	8553
1.27	15.24	4.819	2164	7.228	3246	9.638	4327	14.457	6491	19.276	8655
1.28	15.36	4.876	2189	7.314	3284	9.752	4379	14.628	6568	19.504	8757
1.29	15.48	4.933	2215	7.400	3323	9.866	4430	14.800	6645	19.733	8860
1.3	15.6	4.991	2241	7.486	3361	9.981	4482	14.972	6722	19.963	8963
1.31	15.72	5.048	2267	7.573	3400	10.097	4533	15.145	6800	20.193	9067
1.32	15.84	5.106	2293	7.659	3439	10.213	4585	15.319	6878	20.425	9171
1.33	15.96	5.164	2319	7.747	3478	10.329	4638	15.493	6956	20.658	9275
1.34	16.08	5.223	2345	7.834	3518	10.446	4690	15.668	7035	20.891	9380
1.35	16.2	5.281	2371	7.922	3557	10.563	4743	15.844	7114	21.125	9485
1.36	16.32	5.340	2398	8.010	3597	10.680	4795	16.020	7193	21.361	9591
1.37	16.44	5.399	2424	8.099	3636	10.798	4848	16.197	7273	21.597	9697
1.38	16.56	5.458	2451	8.188	3676	10.917	4902	16.375	7352	21.833	9803
1.39	16.68	5.518	2477	8.277	3716	11.036	4955	16.553	7432	22.071	9910
1.4	16.8	5.577	2504	8.366	3756	11.155	5009	16.732	7513	22.310	10017
1.41	16.92	5.637	2531	8.456	3797	11.275	5062	16.912	7593	22.549	10125
1.42	17.04	5.697	2558	8.546	3837	11.395	5116	17.092	7674	22.790	10233
1.43	17.16	5.758	2585	8.637	3878	11.515	5170	17.273	7756	23.031	10341
1.44	17.28	5.818	2612	8.727	3919	11.636	5225	17.455	7837	23.273	10449
1.45	17.4	5.879	2640	8.818	3959	11.758	5279	17.637	7919	23.516	10558
1.46	17.52	5.940	2667	8.910	4000	11.880	5334	17.819	8001	23.759	10668
1.47	17.64	6.001	2694	9.001	4042	12.002	5389	18.003	8083	24.004	10778
1.48	17.76	6.062	2722	9.093	4083	12.125	5444	18.187	8166	24.249	10888
1.49	17.88	6.124	2750	9.186	4124	12.248	5499	18.371	8249	24.495	10998
1.5	18	6.186	2777	9.278	4166	12.371	5555	18.557	8332	24.742	11109

Float Method

Description

The float method of flow measurement involves placing a float in the water of a ditch, measuring the time it takes to travel a predetermined distance, and then calculating the water velocity from these measurements. The water velocity is then multiplied by the cross-sectional area of the ditch to obtain the flow rate. This may be the most appropriate method for measuring flow rates in temporary earth-lined ditches that are installed along the head of the field each year.

Measuring Flow Rates

The procedure is as follows.

1. Select a straight length of ditch about 100 feet long for the measurement section.
2. Place the float just upstream of the start of the section. When the float reaches the starting point, start measuring the time with a stopwatch. Stop recording the time when the float reaches the end of the section.
3. Calculate the surface water velocity (V_W) using the following formula:

$$V_w \text{ (feet per minute)} = \text{length (feet)} \div \text{travel time (minutes)}$$

4. Multiply this calculated surface water velocity by the appropriate coefficient (C) to calculate the mean or average water velocity (V_{WC}) (see the section "Coefficients, below"):

$$V_{wc} = C \times V_w$$

5. Calculate the cross-sectional area (A_W) of the flow in the ditch (see the section "Cross-Sectional Area," below).
6. Calculate the flow rate, Q, using the following formula:

$$Q \text{ (gallons per minute)} = V_{wc} \text{ (feet per minute)}$$
$$\times A_w \text{ (square feet)} \times 7.48 \text{ gallons per cubic foot}$$

Cross-Sectional Area

To use the float method, the cross-sectional area of the flow must be determined. A trapezoidal shape can be assumed for many earth-lined ditches. The cross-sectional area is determined by measuring the width of the bottom of the ditch (*B*), the depth of water (*D*), and the width of the water surface (*W*) (fig. 37). The unit of measurement is feet. The cross-sectional area of a trapezoidal shape in square feet can be calculated by

$$A_w = ½ \times D \times (W + B)$$

For a V-shaped ditch, the cross-sectional area is calculated by

$$A_w = ½ \times W \times D$$

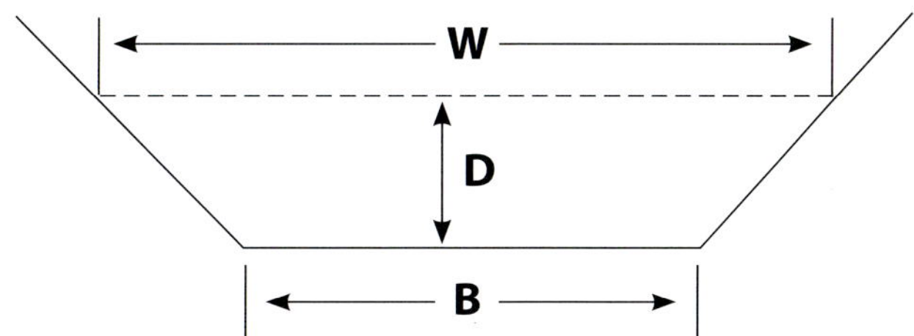

Figure 37. Cross-sectional shape of a trapezoidal ditch showing the width of the bottom of the ditch (B), the depth of water (D), and the width of the water surface (W).

For a ditch that is more or less U-shaped, the cross-sectional area is calculated by

$$A_w = W \times D \times 3.1416 \div 4$$

Coefficients

The water velocity across the ditch can vary, with smaller velocities at the surface and near the ditch bottom and sides. Thus, the surface water velocity must be multiplied by a coefficient (table 6) to calculate the mean or average water velocity in the ditch. The coefficient depends on the depth of water.

Floats

Floats should be partially submerged to reduce any wind effects; items such as oranges, partially filled plastic bottles, or specifically designed rods may be used. Floats that remain completely on the surface such as corks and ping pong balls should be avoided.

Considerations

- Measurements should be made on windless days to prevent the float travel time from being affected by wind.
- The cross-sectional shape should be relatively uniform along the measurement section.
- The measurement section should be free of obstructions, turn-out structures, and excessive vegetative growth in the water.
- The average of several individual measurements of the water velocity is recommended.

Accuracy

The float method provides a rough estimate of the flow rate, with an error of 10 to 20 percent. There can be substantial error in the channel or cross-sectional area determination, especially in irregularly shaped channels.

Table 6. Coefficients for the float method

Average depth (feet)	Coefficient
1	0.66
2	0.68
3	0.70
4	0.72
5	0.74
6	0.76

Source: Water Measurement Manual, 2001.

Current Meters

Description

An approximation of flow rate in a ditch or channel can be made by measuring the cross-sectional area of the channel and the water velocity through the cross-sectional area. The water velocity is measured at selected points using a current meter. A number of current meter devices are available for taking velocity measurements in on-farm ditches. Mechanical current meters use anemometer cups or propellers that rotate in the water. Some of these require the operator to count the number of audible clicks per unit time to determine flow velocity, while others provide a direct velocity readout. Hand-held Doppler and electromagnetic velocity meters can also be used to take point velocity measurements.

Measuring Flow Rates

To measure the flow rate using a current meter, the channel must be divided into a series of cross-sections (fig. 38 and 39). Select a straight section of channel with a representative cross-sectional shape (this is often not difficult with ditches formed by farm implements). Use a sturdy extension ladder with a board on the rungs as a bridge across the ditch. Lay a measuring tape along the board to mark the measurement locations. Take measurements of flow depth (*d*) and velocity (*V*) at 2-foot increments (*L*) across the channel. A plastic surveying rod works well to measure depth, although a carpenter's rule or tape measure will also work. Record all vertical and horizontal measurements.

If *d* is less than 2 feet at any location *L*, take one velocity measurement at the "six-tenths depth" (0.6 times *d*) for each location. If *d* is greater than 2 feet at every location *L*, take two velocity measurements at each location: one at 0.2 times *d*, and the second at 0.8 times *d*. Carefully record all measurements.

The average flow velocity ($\bar{V}$) at each vertical location *L* is the average of the two measurements taken, or the single measurement if the six-tenths depth method was used.

For trapezoidal and triangular channels, the flow rate is

$$\left\{\left[\frac{\bar{V}_1}{2}\times\frac{d_1}{2}\right]L1\right\}+\left\{\left[\frac{\bar{V}_1+\bar{V}_2}{2}\left(\frac{d_1+d_2}{2}\right)\right](L_2-L_1)\right\}+\left\{\left[\left(\frac{\bar{V}_2+\bar{V}_3}{2}\right)\left(\frac{d_2+d_3}{2}\right)\right](L_3-L_2)\right\}$$
$$+\left\{\left[\frac{\bar{V}_3}{2}\times\frac{d_3}{2}\right](L_4-L_3)\right\}$$

If the channel is wide, take additional measurements of *V*, *L*, and *d* to more accurately describe the channel flow rate. Keep the measurement units consistent. For example, if *L* and *d* are in feet and *V* is in feet per second, the flow rate is in cubic feet per second (cfs).

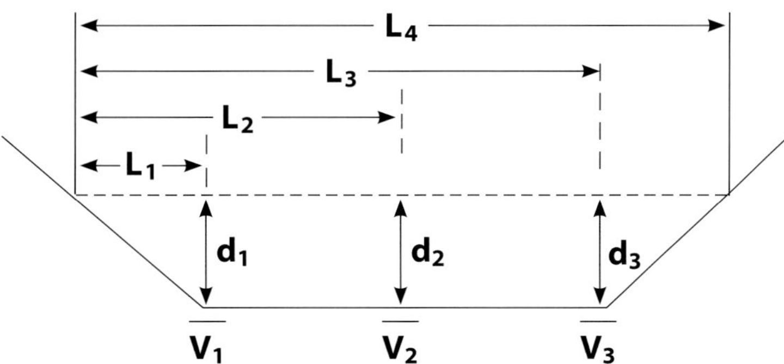

Figure 38. Current meter measurements in a trapezoidal channel: width (L), depth (d), and velocity (V) measurements.

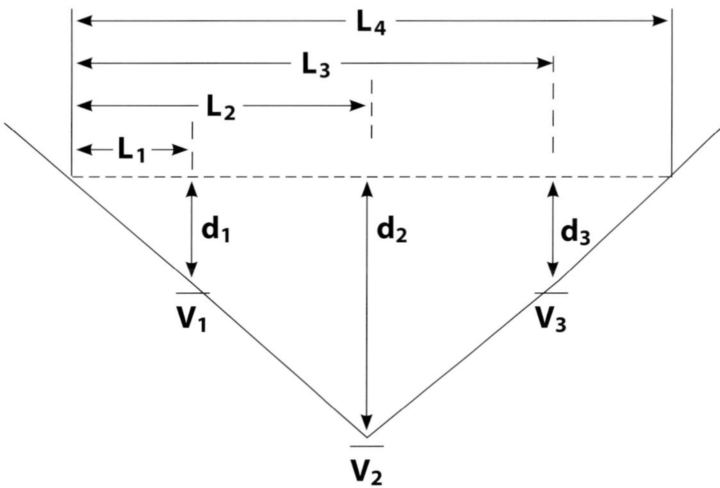

Figure 39. Current meter measurements in a triangular channel: width (L), depth (d), and velocity (V) measurements.

Measuring Siphon Flow Rates

Description

Measuring flow rates in siphons that deliver water to furrows or border checks requires measuring the siphon diameter and the head. The head is the difference between the elevation of the water surface in the channel and the elevation of the water at the siphon discharge in the field (fig. 40). The head is easily measured with a surveying level, but if one is not available, use a commercially available construction water level or a carpenter's level and tape measure (fig. 41).

Measuring Flow Rates

Individual siphon flow rates are determined using figure 42. The total flow rate to the irrigated area is

$$\text{total flow rate to irrigated area (gpm or cfs)} = \text{siphon flow rate (from fig. 42, gpm or cfs)} \times \text{number of siphons operating}$$

Note that 1 cfs equals 449 gpm. It is often convenient to convert the total flow rate of the irrigated area (gpm or cfs) to an irrigated area application rate (in/hr) using the following formulas:

$$\text{application rate (in/hr)} = 0.0022 \times \text{total flow rate (gpm)} \div \text{irrigated area (acres)}$$

or

$$\text{application rate (in/hr)} = \text{total flow rate (cfs)} \div \text{irrigated area (acres)}$$

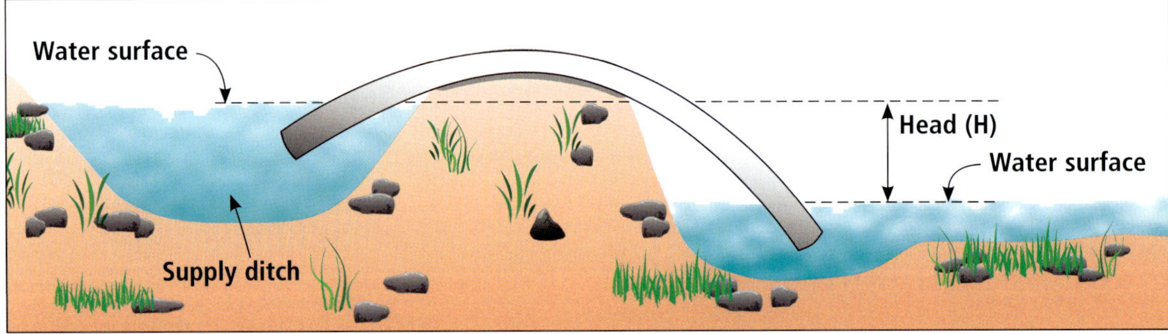

Figure 40. Measuring the head of siphons. The head is the difference between the elevation of the water surface in the ditch and the water surface in the field.

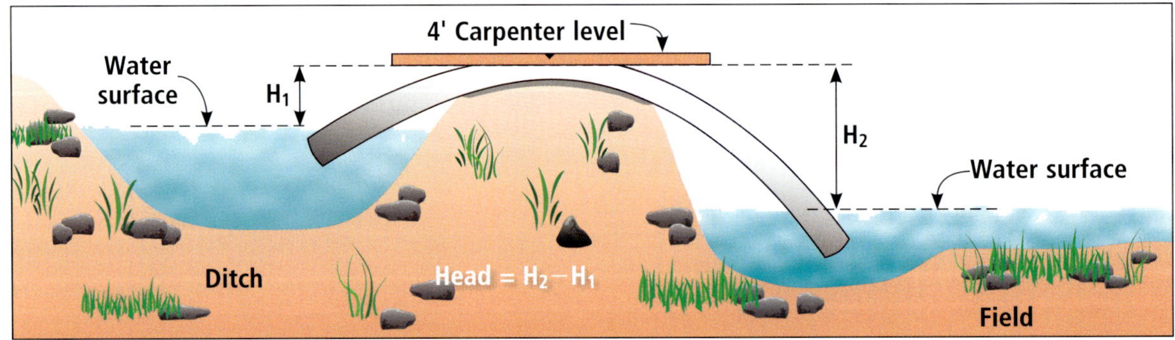

Figure 41. Measuring the head of siphons using a carpenter level.

Accuracy

Two sources of error can occur when using the siphon flow rate to determine the field flow rate. First, the diameter of siphons can vary. For example, the siphon being measured in the field might be slightly narrower or wider than the 2-inch siphon used to develop the data in figure 43. Second, in furrow irrigation, each furrow is usually a separate channel, and the downstream water height in each furrow may depend on the siphon placement. To minimize this source of error, determine an average downstream water surface elevation from measurements made in several furrows. Calculations of siphon flow rate in border irrigations are less susceptible to variations in downstream surface elevation compared with those made under furrow irrigation.

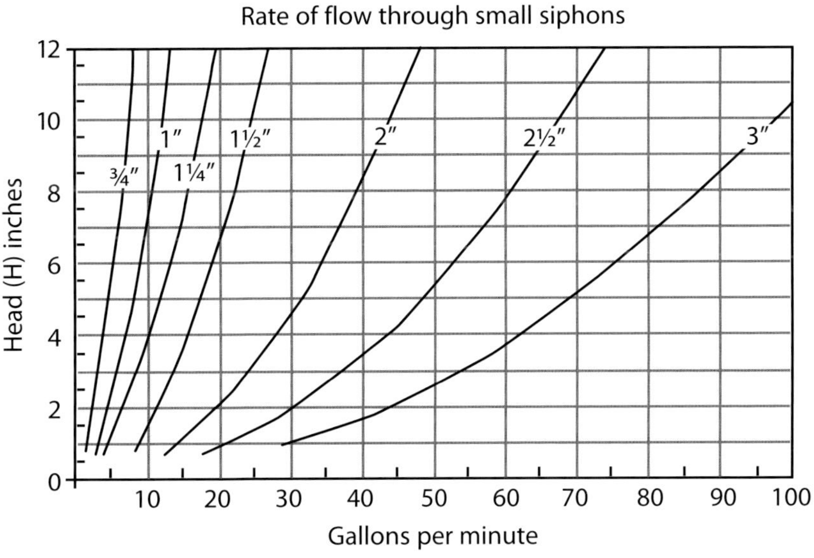

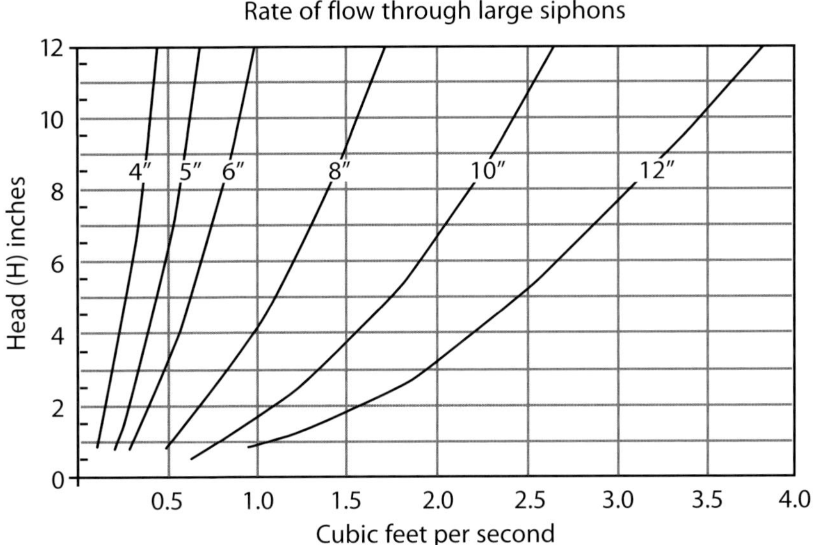

Figure 42. Flow rate figures for small and large siphons. To determine the flow rate, start at the left, move horizontally from the point of measured head to the size of siphon, then move vertically down to find the flow rate. *Source*: Scott and Houston 1977.

Volumetric Measurements

Description

Volumetric measurement of the flow rate involves measuring the time required for the discharge of a volume of water into a container. This method is sometimes called the "bucket and stopwatch" method. The flow rate (Q) is then calculated using the following equation.

$$Q = \text{volume} \div \text{time}$$

The unit used for the flow rate depend on the units used for the volume and time measurements.

Measuring Flow Rates

The equipment required for this method consists of a container and a stopwatch. The procedure involves measuring the time to fill the container with water. However, some method of determining the volume of water in the container must be used. A graduated cylinder can be used to measure it. The water can also be caught in a 5-gallon bucket and either poured into a graduated cylinder or weighed to measure the volume of water. The weight of the water (total weight minus bucket weight) can be converted to a volume using the conversion factor of 1 gallon of water equals 8.33 pounds.

This method is appropriate for small flow rates, such as the individual flow rates of gated pipe or sprinkler nozzles. It is particular appropriate for measuring the discharge rates of sprinkler nozzles. Place a short section of garden hose over the nozzle and direct the flow into a graduated cylinder. Record the volume caught and the elapsed time between the start and end of the flow and calculate the rate using the above equation.